INTRODUCTION

- We have used colour sparingly in this guide and never purely for the sake of it.
  The text remains largely in black and white so that pages are spacious and uncluttered.

- The summary questions are in black and white and are in the same multiple choice format as pupils will face in their modular tests.

- It's also important to realise that while the book is now 72 pages, the amount of material required to be revised remains the same as in our earlier editions. Although we have expanded some explanations and examples, the same care has been taken to ensure that these notes are distilled to the "essentials".

- Although the guide mostly follows the syllabus order a few topics have been sympathetically relocated for better understanding. However the contents page links each page to its syllabus reference number so that pages can be easily cross-referenced to their syllabus location.

- Apart from these innovations, the same principles still apply ...

  ... i.e. the guide is written SPECIFICALLY FOR YOUR SYLLABUS and contains ...

  ... ALL THE INFORMATION THAT NEEDS TO BE KNOWN and just as importantly ...

  ... NO INFORMATION THAT ISN'T REQUIRED BY THE SYLLABUS.

- Those sections of the guide which are OUTLINED IN RED are the areas of the syllabus, previously tested by modular test, which may be TESTED AGAIN IN YOUR TERMINAL EXAMINATION. These are the areas which must be revised again thoroughly towards the end of year 11.

- The second volume of this guide concentrates on the six modules which will be tested in the TERMINAL EXAMINATION. It is available from the address on the opposite page.

- The KEY POINTS are summarised at the bottom of each page.

CONTENTS

Covered in Class | Revised | Revised | Page No.

Humans as Organisms

5   Life Processes and Organisation of Life. (1.1)
6   Animal Cells and Diffusion. (1.1/1.6)
7   The Digestive System. (1.2)
8   Digestive Enzymes and Diet. (1.2)
9   Breathing - Structure of the Lungs. (1.3)
10   Respiration. (1.3)
11   Circulation I - The Blood. (1.4)
12   Circulation II - The Heart. (1.4)
13   Circulation III - The Blood Vessels. (1.4)
14   Disease. (1.5)
15   Summary Questions.
16   Summary Questions.

Maintenance of Life

17   Plant Cells. (2.1)
18   Plant Structure and Photosynthesis. (2.2)
19   Factors affecting Photosynthesis. (2.2)
20   Transpiration. (2.3)
21   Water Loss and Wilting. (2.3)
22   Plant Hormones. (2.4)
23   The Nervous System and The Eye. (2.5)
24   Homeostasis. (2.6)
25   Drugs. (2.7)
26   Summary Questions.
27   Summary Questions.

Metals

28   Characteristics and Uses of Metals. (5.1)
29   Reactivity of Metals I - The Reactivity Series. (5.2)
30   Reactivity of Metals II - Displacement Reactions. (5.2)
31   Extracting Metals from their Ores I. (5.3)
32   Extracting Metals from their Ores II. (5.3)
33   Neutralisation and Salt Formation I. (5.4)
34   Neutralisation and Salt Formation II. (5.4)
35   Summary Questions.
36   Summary Questions.

Covered in Class Revised Revised Page No.

Earth Materials

37 Uses of Limestone. (6.1)
38 Rock Types and Rock Formation. (6.2)
39 Rock Cycle. (6.2)
40 Oil I - What it is and how it's formed. (6.3)
41 Oil II - Fractional Distillation and Cracking. (6.3)
42 Effect of Burning on the Atmosphere I - Global Warming. (6.4)
43 Effect of Burning on the Atmosphere II - Acid Rain. (6.4)
44 Earth Structure and Movement. (6.5)
45 Tectonic Plates. (6.5)
46 Summary Questions.
47 Summary Questions.

Energy

48 Thermal Energy Transfer I - Conduction, Convection and Radiation. (9.1)
49 Thermal Energy Transfer II - Insulation. (9.1)
50 Electrical Energy Transfer and Power. (9.2)
51 Cost of Using Electrical Appliances. (9.2)
52 Gravitational Potential Energy. (9.2)
53 Efficiency of Energy Transfer. (9.3)
54 Energy Resources I - Generation of Electricity. (9.4)
55 Energy Resources II - Non-renewable Resources. (9.4)
56 Energy Resources III - Renewable Resources. (9.4)
57 Summary Questions.
58 Summary Questions.

Electricity

59 Potential Difference (P.d) and Current in Circuits. (10.1)
60 Series and Parallel Circuits. (10.1)
61 Static Electricity. (10.3)
62 Static in Real Life and Electrolysis. (10.3)
63 Mains Electricity I - The 3-Pin Plug. (10.4)
64 Mains Electricity II - Safety Devices. (10.4)
65 Energy and Power in Circuits; (plus Electrical Symbols). (10.5/10.1)
66 Electromagnetic Forces I - Magnets, Electromagnets, and the Motor Effect (10.2)
67 Electromagnetic Forces II - Making things move. (10.2)
68 Generators and Transformers. (10.6)
69 Summary Questions.
70 Summary Questions.

71 Index.
72 Index.

*Numbers in brackets refer to Syllabus reference numbers.

HOW TO USE THIS REVISION GUIDE

- Don't just read! LEARN ACTIVELY!

- Constantly test yourself ... WITHOUT LOOKING AT THE BOOK.

- When you have revised a small sub-section or a diagram, PLACE A BOLD TICK AGAINST IT.
  Similarly, tick the "progress and revision" section of the contents when you have done a page.
  This is great for your self confidence.

- Jot down anything which will help YOU to remember - no matter how trivial it may seem.

- DON'T BE TEMPTED TO HIGHLIGHT SECTIONS WITH DIFFERENT COLOURS.
  TOO MUCH COLOUR REDUCES CLARITY AND CAUSES CONFUSION.
  YOUR EXAM WILL BE IN BLACK AND WHITE!

- These notes are highly refined. Everything you need is here, in a highly organised
  but user-friendly format. Many questions depend only on STRAIGHTFORWARD RECALL
  OF FACTS, so make sure you LEARN THEM.

- Remember, don't throw away this guide at the end of year 10!!
  You will need to revise all the sections OUTLINED IN RED in this volume as well as the SIX modules
  in Volume Two before you take your final exam at the end of year 11.

- THIS IS YOUR BOOK! Use it throughout your course in the ways suggested
  and your revision will be both organised and successful.

SOME IMPORTANT FACTS ABOUT YOUR MODULE TESTS

- You will take 6 module tests over the duration of your course.

- If you miss a test through illness or another reason you may take it again on one of the specified occasions,
  but YOU MAY NOT REPEAT A TEST.

- The specified occasions are EARLY DECEMBER, EARLY MARCH and MID JUNE, on dates
  which are announced each year. You may also do tests in year 11 on the first two of these specified occasions.

- You will be entered for each of these tests at either HIGHER or FOUNDATION TIER. You don't have to stick
  to one tier for all six tests.

- The tests will be entirely MULTIPLE CHOICE and similar in style to the summary questions in this guide.

- There are a maximum of 36 marks for either tier, and in previous tests 18 of the marks have been for
  questions common to both tiers.

HUMANS AS ORGANISMS

THE 7 LIFE PROCESSES

These processes to some extent are common to all life forms ...

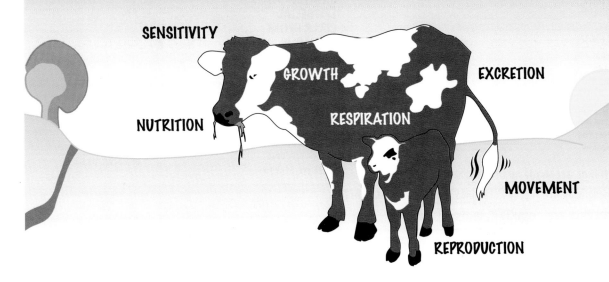

SENSITIVITY
GROWTH
EXCRETION
NUTRITION
RESPIRATION
MOVEMENT
REPRODUCTION

MOVEMENT	Muscles contract and parts of the body move.
REPRODUCTION	All living things need to produce offspring.
SENSITIVITY	The ability to respond to change in the surroundings.
GROWTH	Organisms which are born small need to grow into adult size.
RESPIRATION	Oxygen is combined with food to release energy in cells.
EXCRETION	Waste products from reactions in cells must be removed.
NUTRITION	Food is needed for new body structure and energy.

The initial letters of the 7 processes spell MRS.GREN! Remember this!
Also remember that plants do all these things ... but less obviously!

ORGANISATION OF LIFE

- A group of cells with similar structure and function is called a TISSUE,
  e.g. MUSCLE TISSUE, NERVOUS TISSUE, BLOOD, GLANDULAR and XYLEM TISSUE (in plants).

- Two or more tissues working together is called an ORGAN,
  e.g. BRAIN, KIDNEY, HEART, LUNG.

- Organs can work together with other organs to form ORGAN SYSTEMS,
  e.g. Cardiovascular system (Heart, lungs, blood vessels, blood).

- Several organ systems form an ORGANISM.

CELLS ⟹ TISSUES ⟹ ORGANS ⟹ ORGAN SYSTEMS ⟹ ORGANISM

- The seven life processes are: Movement, Reproduction, Sensitivity, Growth, Respiration, Excretion and Nutrition.
- Life is organised in the following way: Cells ⟶ Tissues ⟶ Organs ⟶ Organ Systems ⟶ Organism.

TYPICAL ANIMAL CELLS

A CHEEK CELL FROM A HUMAN

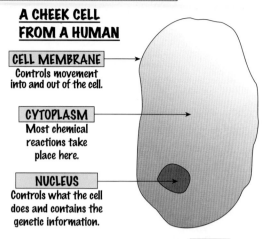

CELL MEMBRANE
Controls movement into and out of the cell.

CYTOPLASM
Most chemical reactions take place here.

NUCLEUS
Controls what the cell does and contains the genetic information.

- All Living things are made up of **CELLS.**
- Some of these cells are highly **SPECIALISED** to do a certain job.
- We say that they are **ADAPTED** to their function.

SPECIALISED ANIMAL CELLS

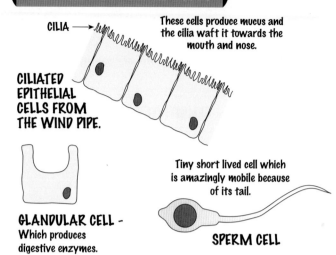

CILIA → These cells produce mucus and the cilia waft it towards the mouth and nose.

CILIATED EPITHELIAL CELLS FROM THE WIND PIPE.

GLANDULAR CELL - Which produces digestive enzymes.

Tiny short lived cell which is amazingly mobile because of its tail.

SPERM CELL

DIFFUSION ...

Diffusion is ... the spreading of a GAS or any substance in SOLUTION ...

... from a HIGHER to a LOWER CONCENTRATION.

The greater the difference in concentration, the faster the rate of diffusion.

Two simple examples are:
- The smell of perfume spreading throughout a room.
- A drop of ink spreading throughout a beaker of water.

TWO EXAMPLES IN HUMANS

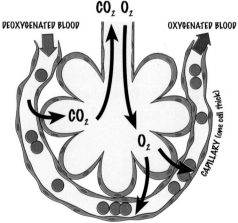

CO_2 O_2

DEOXYGENATED BLOOD

OXYGENATED BLOOD

CO_2

O_2

CAPILLARY (one cell thick)

❶ An alveolus in the lungs.

Here, ...
- ... Oxygen diffuses from the alveolus into the blood ...
- ... and carbon dioxide diffuses from the blood into the alveolus.

i.e. Both substances are going from a higher to a lower concentration.

❷ Muscle cells in the body.
Here, ...
- ... Oxygen and sugar (glucose) diffuse from the blood into the cells ...

- ... and carbon dioxide, water and waste diffuse from the cells into the blood (ions may also diffuse either way).

i.e. All substances are moving from a higher to a lower concentration.

MUSCLE CELLS

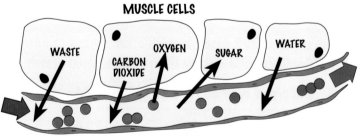

WASTE

CARBON DIOXIDE

OXYGEN

SUGAR

WATER

A CAPILLARY VESSEL (one cell thick)

- The key labels are: Cell Membrane, Cytoplasm, Nucleus.
- Diffusion is the spreading of a gas or any substance in solution from a higher to a lower concentration.

The DIGESTIVE SYSTEM is really made up of a long MUSCULAR TUBE in which ENZYMES speed up (catalyse) the breakdown of LARGE INSOLUBLE MOLECULES into SMALLER SOLUBLE MOLECULES so that they can pass through the walls of the small intestine and into the bloodstream.

THE HUMAN DIGESTIVE SYSTEM

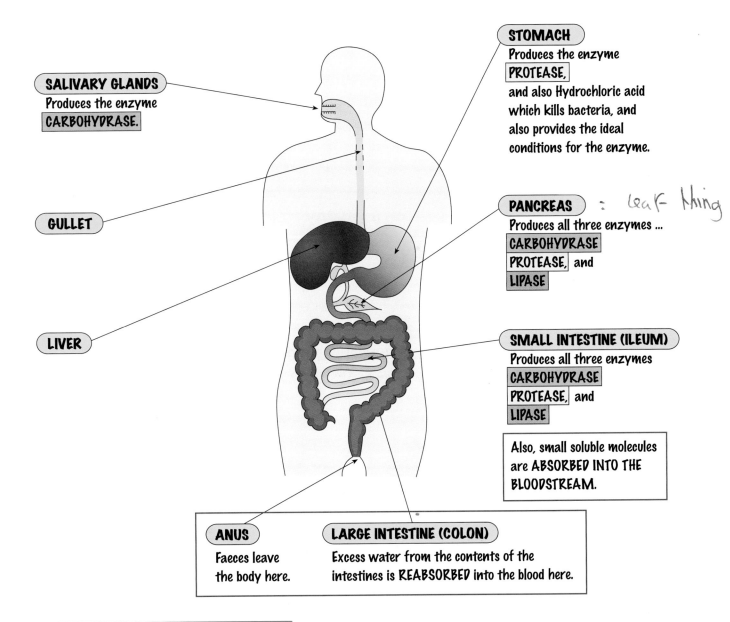

SALIVARY GLANDS
Produces the enzyme
CARBOHYDRASE.

GULLET

LIVER

STOMACH
Produces the enzyme
PROTEASE,
and also Hydrochloric acid
which kills bacteria, and
also provides the ideal
conditions for the enzyme.

PANCREAS : leaf thing
Produces all three enzymes ...
CARBOHYDRASE
PROTEASE, and
LIPASE

SMALL INTESTINE (ILEUM)
Produces all three enzymes
CARBOHYDRASE
PROTEASE, and
LIPASE

Also, small soluble molecules
are ABSORBED INTO THE
BLOODSTREAM.

ANUS
Faeces leave
the body here.

LARGE INTESTINE (COLON)
Excess water from the contents of the
intestines is REABSORBED into the blood here.

TISSUES OF THE DIGESTIVE SYSTEM

Two outer layers of MUSCLE TISSUE
to push the food along.

An inner, folded layer of GLANDULAR TISSUE
to make enzymes to catalyse the
breakdown of the food.

A SECTION OF THE SMALL
INTESTINE WHICH JUST
HAPPENS TO BE TYPICAL OF
THE DIGESTIVE SYSTEM.

- The mouth produces just Carbohydrase, and the stomach just Protease.
- The Pancreas and the Small Intestine produce Carbohydrase, Protease and Lipase.

HUMAN DIET Three main food substances are needed by humans ...

WHAT THEY ARE	WHAT WE USE THEM FOR	WHERE WE GET THEM

PROTEINS

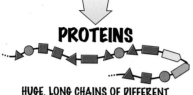

HUGE, LONG CHAINS OF DIFFERENT AMINO ACIDS.

CARBOHYDRATES

HUGE, LONG CHAINS OF IDENTICAL SUGAR MOLECULES (STARCH) OR,

INDIVIDUAL SUGAR MOLECULES (GLUCOSE).

FATS

FAT MOLECULES CONSISTING OF THREE FATTY ACIDS ATTACHED TO A MOLECULE OF GLYCEROL

GROWTH AND REPAIR, AND REPLACEMENT OF CELLS

IMMEDIATE ENERGY
THROUGH RESPIRATION.

STORED ENERGY
AND HEAT INSULATION AND MAKING CELL MEMBRANES.

MEAT   FISH   EGGS   BEANS (pulses)

POTATOES (root vegetables)   BREAD   FRUITS   CEREALS

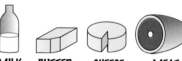
MILK   BUTTER and MARGARINE   CHEESE   MEAT

THE ENZYMES OF THE DIGESTIVE SYSTEM

Because Proteins, Carbohydrates and Fats are LARGE INSOLUBLE MOLECULES, they must be broken down into SMALLER SOLUBLE MOLECULES before they can be ABSORBED. ENZYMES DO THIS, AND THERE ARE THREE OF THEM.

SMALL INTESTINE

PROTEASE ENZYMES catalyse the breakdown of PROTEINS into AMINO ACIDS.

PROTEINS

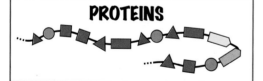

AMINO ACIDS

CARBOHYDRASE ENZYMES catalyse the breakdown of STARCH (carbohydrate) into SUGARS.

CARBOHYDRATES (e.g. Starch)

SUGARS

LIPASE ENZYMES catalyse the breakdown of FATS into FATTY ACIDS and GLYCEROL.

FATS

FATTY ACIDS & GLYCEROL

ABSORPTION INTO BLOOD STREAM

LOOK CAREFULLY AT THE PREVIOUS PAGE TO SEE WHERE THESE ENZYMES ARE MADE.

- Proteins (for growth and repair) are converted to amino acids by Proteases. • Carbohydrates (for immediate energy) are converted to sugars by Carbohydrases. • Fats (for stored energy) are converted to fatty acids and glycerol by Lipases.

The lungs are designed to enable gases (oxygen and carbon dioxide) to be efficiently exchanged. The alveoli themselves provide a massive surface area for gas exchange and are a good example of a specialised exchange surface.

CONTENTS OF THE THORAX (CHEST CAVITY)

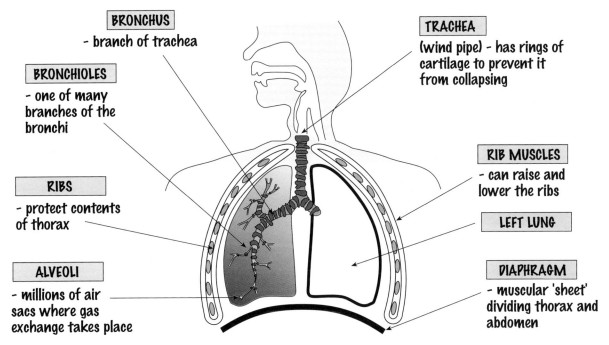

BRONCHUS
- branch of trachea

BRONCHIOLES
- one of many branches of the bronchi

RIBS
- protect contents of thorax

ALVEOLI
- millions of air sacs where gas exchange takes place

TRACHEA
(wind pipe) - has rings of cartilage to prevent it from collapsing

RIB MUSCLES
- can raise and lower the ribs

LEFT LUNG

DIAPHRAGM
- muscular 'sheet' dividing thorax and abdomen

- The RIBCAGE protects the contents of the THORAX i.e. the HEART and LUNGS.

DIFFUSION OF OXYGEN AND CARBON DIOXIDE IN THE ALVEOLI

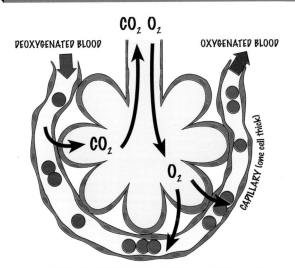

CO_2 O_2

DEOXYGENATED BLOOD

OXYGENATED BLOOD

CO₂

O₂

CAPILLARY (one cell thick)

A SINGLE ALVEOLUS AND A CAPILLARY.

- The TRACHEA divides into two tubes ...
- ... called the BRONCHI which divide again, several times ...
- ... to form the BRONCHIOLES which continue to divide ...
- ... until they end as air sacs called ALVEOLI (There are millions of these) ...
- ... which are very close to the blood CAPILLARIES.

and here, at this GAS EXCHANGE SURFACE ...
- CARBON DIOXIDE diffuses from the BLOOD into the ALVEOLI.
- OXYGEN diffuses from the ALVEOLI into the blood.

This means the blood has swapped its CARBON DIOXIDE for OXYGEN and is now OXYGENATED.

DIFFUSION

All it means is ...
- ... the SPREADING OF A GAS ...
- ... OR ANY SUBSTANCE IN SOLUTION ...
- ... from a HIGHER to a LOWER concentration.

- Trachea, Bronchus, Bronchioles, Alveoli, Ribs, Rib Muscles and Diaphragm are the key labels.
- Carbon Dioxide diffuses from the blood into the alveoli. Oxygen diffuses from the alveoli into the blood.

AEROBIC RESPIRATION

AEROBIC RESPIRATION is THE RELEASE OF <u>ENERGY</u> FROM THE <u>BREAKDOWN OF GLUCOSE</u> ...

- ... BY COMBINING IT WITH <u>OXYGEN</u> ... INSIDE LIVING CELLS.

THE EQUATION:

$$\text{GLUCOSE + OXYGEN} \Longrightarrow \text{CARBON DIOXIDE + WATER + ENERGY}$$

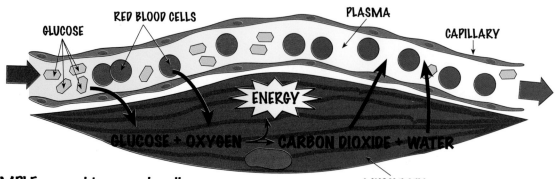

EXAMPLE: a working muscle cell

- The energy is contained INSIDE THE GLUCOSE molecule, and comes from the food we eat.
- The oxygen has come from the atmosphere, via the lungs.

ANAEROBIC RESPIRATION

In a nutshell, this means:-

- THE RELEASE OF A LITTLE <u>BIT OF ENERGY</u> FROM THE <u>INCOMPLETE BREAKDOWN OF GLUCOSE</u>.
- .... in THE <u>ABSENCE OF OXYGEN</u> ...
- ... INSIDE LIVING CELLS.

THE EQUATION:

$$\text{GLUCOSE} \Longrightarrow \text{A BIT OF ENERGY + LACTIC ACID}$$

- This happens when the muscles are working so hard that ...
- ... the lungs and bloodstream can't deliver enough OXYGEN, to respire the available glucose aerobically.
- Therefore the GLUCOSE can only be partly broken down, releasing a much smaller amount of energy ...
- ... and LACTIC ACID as a waste product. It can only operate for a short time.

WHAT CELLS DO WITH THEIR ENERGY

The energy released during respiration is used in the following ways ...

- To build larger MOLECULES using smaller ones.
- To enable MUSCLES to contract.
- To MAINTAIN TEMPERATURE in colder surroundings.

They all begin with **M**.
(Remember the "M's")

- Aerobically: GLUCOSE + OXYGEN ⟶ CARBON DIOXIDE + WATER + ENERGY
- Anaerobically: GLUCOSE ⟶ A BIT OF ENERGY + LACTIC ACID
- Cells use energy for making molecules, muscle contraction and for maintaining temperature.

THE CIRCULATORY SYSTEM

The CIRCULATORY SYSTEM is a system for transporting substances around the body. Things which are taken into the body need to be transported to the cells, and substances from the cells need to be transported to where they are removed from the body.

The CIRCULATORY SYSTEM can be divided into ...

> The Blood
> The Heart, and ...
> The Blood Vessels (Arteries, veins and capillaries)

THE BLOOD

If blood is allowed to stand without clotting, it separates out into layers due to gravity ...

- ... these layers represent:-　　the RED CELLS and PLATELETS.
　　　　　　　　　　　　　　　　the WHITE CELLS and ...
　　　　　　　　　　　　　　　　... the PLASMA.

← PLASMA
← WHITE CELLS
← RED CELLS AND PLATELETS

Briefly, these components of the blood have the following functions

RED CELLS	-	transport OXYGEN
PLATELETS	-	help the blood to CLOT
WHITE CELLS	-	DEFEND against microbes

PLASMA
　　transports CARBON DIOXIDE
　　transports UREA
　　transports FOOD

---

1. The Red Blood Cells ...

... transport OXYGEN from the lungs to the organs.
They have NO NUCLEUS so that they can contain lots of HAEMOGLOBIN, (a red pigment which can carry oxygen).
In the lungs HAEMOGLOBIN combines with OXYGEN to form OXYHAEMOGLOBIN.
In other organs OXYHAEMOGLOBIN splits up into HAEMOGLOBIN plus OXYGEN.

---

2. The Platelets ...

... are tiny pieces of cells which have no nucleus.
They are an important factor in helping the blood to clot when a blood vessel has been damaged.

---

3. The White Blood Cells ...

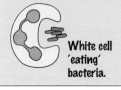

White cell 'eating' bacteria.

White cell making antibodies to make bacteria stick together.

... have a NUCLEUS which may be quite variable in shape.
They help to DEFEND the body against invading MICROBES (see also p.14).

---

4. The Plasma ...

... the cells above are suspended in the plasma which is a straw-coloured liquid.
Plasma transports ...

❶ Carbon dioxide from the organs to the lungs.
❷ Soluble products of digestion (e.g. glucose) from the small intestine to the organs.
❸ Urea from the liver to the kidneys.

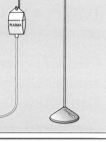

○　• The six main jobs of blood can be summarised as Food, Oxygen, Waste, Heat, Defence and Clotting. • Red Blood Cells carry oxygen. White Blood Cells are for defence. Platelets are for clotting. Plasma carries food, waste and carbon dioxide.

The heart is the pump which forces blood through the blood vessels and around the circulatory system.

THE PATHWAY OF THE CIRCULATION

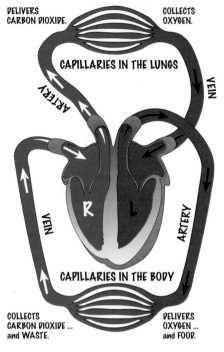

DELIVERS CARBON DIOXIDE.

COLLECTS OXYGEN.

CAPILLARIES IN THE LUNGS

ARTERY

VEIN

R   L

VEIN

ARTERY

CAPILLARIES IN THE BODY

COLLECTS CARBON DIOXIDE ... and WASTE.

DELIVERS OXYGEN ... and FOOD.

■ Blood low in oxygen (DEOXYGENATED)

■ Blood rich in oxygen (OXYGENATED)

There are TWO SEPARATE CIRCULATION SYSTEMS, ...

> One which carries blood from the HEART to the LUNGS and then back to the HEART ...

> ... and one which carries blood from the HEART to ALL OTHER PARTS OF THE BODY and then back to the HEART.

- This means that blood flows around a 'figure of eight' circuit and passes through the heart TWICE on each circuit.
- Blood travels AWAY from the heart through ARTERIES, ...
- ... and returns to the heart through VEINS.

> The RIGHT SIDE of the heart pumps blood which is LOW IN OXYGEN to the LUNGS, to pick up OXYGEN.

> The LEFT SIDE of the heart pumps blood which is RICH IN OXYGEN and delivers it to all other parts of the BODY.

THE DIFFERENT PARTS OF THE HEART IN MORE DETAIL

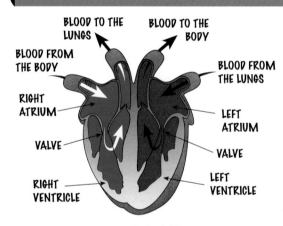

BLOOD TO THE LUNGS

BLOOD TO THE BODY

BLOOD FROM THE BODY

BLOOD FROM THE LUNGS

RIGHT ATRIUM

LEFT ATRIUM

VALVE

VALVE

RIGHT VENTRICLE

LEFT VENTRICLE

- Most of the wall of the heart is made of MUSCLE.

- ATRIA are the smaller, <u>less muscular</u> upper chambers, which receive blood coming back to the heart through VEINS.

- VENTRICLES are the larger, <u>more muscular</u> lower chambers. The LEFT is more muscular than the right since, it has to pump blood around the whole body.

- VALVES make sure that the blood flows in the right direction, and can't flow backwards

HOW THE HEART PUMPS BLOOD

- When the heart muscle relaxes, blood flows into the ATRIA through VEINS from the LUNGS and rest of the BODY.

- The ATRIA then contract, squeezing blood into the VENTRICLES.

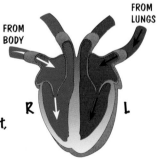

FROM BODY

FROM LUNGS

R   L

- When the VENTRICLES contract, blood is forced, under HIGH PRESSURE into the two ARTERIES which carry blood to the BODY and LUNGS.

- The heart muscle now relaxes and the whole process starts again.

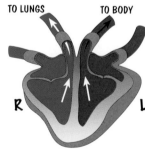

TO LUNGS

TO BODY

R   L

- Key labels are Left and Right Atria, Left and Right Ventricles, Valves, Arteries, Veins and Capillaries.
- The right side of the heart pumps deoxygenated blood to the lungs.
- The left side of the heart pumps oxygenated blood to the body.

There are three types of blood vessels ... ARTERIES, VEINS and CAPILLARIES.
They form the "plumbing" of the circulatory system.

ARTERIES

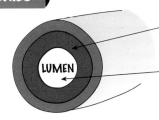

- Thick wall containing ELASTIC and MUSCLE fibres to cope with the much higher pressure in these vessels.
- Much smaller lumen compared to the thickness of the wall.
- No valves.
- Carry blood <u>away</u> from the heart.

VEINS

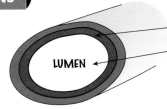

- Thinner wall containing <u>LESS</u> ELASTIC and MUSCLE fibres.
- Much bigger lumen compared to the thickness of the wall.
- Have valves to prevent backflow of blood.
- Carry blood <u>towards</u> the heart.

VALVE

CAPILLARIES

- Narrow, thin-walled vessels, just ONE CELL THICK.
- Microscopic - (too small to see without a microscope).
- Exchange of substances between cells and blood <u>ONLY</u> takes place here.
- Connect arteries to veins.

EXCHANGE OF SUBSTANCES AT THE CAPILLARIES

The heart and blood vessels provide a route around the body, but it is the BLOOD which actually TRANSPORTS substances. Exchange of substances between the blood and the body tissues can <u>ONLY</u> OCCUR IN CAPILLARIES.

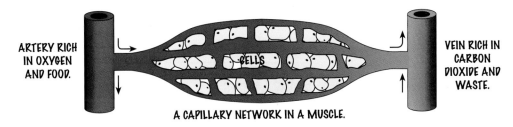

ARTERY RICH IN OXYGEN AND FOOD.

CELLS

VEIN RICH IN CARBON DIOXIDE AND WASTE.

A CAPILLARY NETWORK IN A MUSCLE.

MUSCLE CELLS

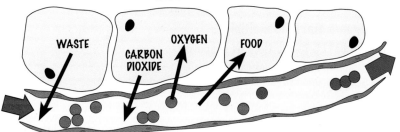

FOOD and OXYGEN DIFFUSE from the blood into the cells.

WASTE   OXYGEN   FOOD
CARBON DIOXIDE

WASTE and CARBON DIOXIDE DIFFUSE from the cells into the blood.

A CAPILLARY VESSEL
(one cell thick)

- Arteries are thick-walled and elastic and carry blood away from the heart.
- Veins are thinner-walled and less elastic, have valves and carry blood back to the heart.
- Capillaries are one cell thick and are the only vessels through which exchange of substances can occur.

BACTERIA AND VIRUSES

These are the two main types of microbe which may affect health.

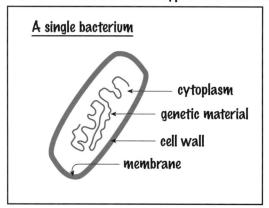

A single bacterium
- cytoplasm
- genetic material
- cell wall
- membrane

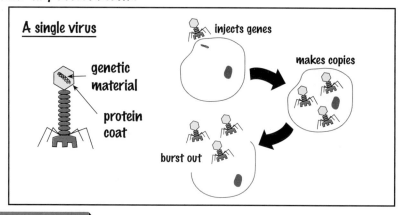

A single virus
- genetic material
- protein coat
- injects genes
- makes copies
- burst out

A COMPARISON BETWEEN BACTERIA AND VIRUSES

BACTERIA	VIRUSES
Consist of CYTOPLASM and a MEMBRANE surrounded by a CELL WALL.	Have a simple PROTEIN COAT. No membrane or cell wall.
The genetic material is NOT contained within a NUCLEUS.	The genetic material is NOT contained within a NUCLEUS.
Very small.	Even smaller.
Reproduce very quickly.	Reproduce very quickly - BUT ONLY INSIDE LIVING CELLS, WHICH ARE THEN DAMAGED.
Can produce TOXINS (poisons) which make us feel ill.	Can produce TOXINS (poisons) which make us feel ill.
Responsible for diseases such as, TETANUS, CHOLERA, TUBERCULOSIS.	Responsible for diseases such as, COLDS, FLU, MEASLES, POLIO.

Diseases are more likely to occur if large numbers of microbes invade the body due to UNHYGENIC CONDITIONS or CONTACT WITH INFECTED PEOPLE.

OUR DEFENCE AGAINST MICROBES

1. The blood produces CLOTS that seal cuts

2. The BREATHING ORGANS produce a STICKY, LIQUID MUCUS, which covers the lining of these organs and traps microbes.

3. The SKIN acts as a barrier to invading microbes

4. The White Cells
- A microbe invades the body and starts to multiply ...
- ... causing the body's WHITE CELLS to multiply in response.
- They defend by
    - (A) INGESTING MICROBES.
    - (B) PRODUCING ANTIBODIES TO DESTROY PARTICULAR MICROBES.
    - (C) PRODUCING ANTITOXINS TO NEUTRALISE TOXINS PRODUCED BY THE MICROBES.
- Once the white cells have been exposed to a particular microbe ...
- ... they can produce the relevant antibodies much quicker next time ...
- ... meaning that the individual now has NATURAL IMMUNITY to the disease.

A white cell ingesting (eating) microbes.

- Bacteria look like cells but have no distinct nucleus. Viruses are just a protein coat surrounding a few genes. They can only reproduce <u>inside</u> living cells. • The body's defence consists of Clotting, Mucus, Skin and the White Cells.

1. The table below is about the digestive system. Match words from the list with each of the numbers 1-4 in the table.
   A. LARGE INTESTINE.
   B. SMALL INTESTINE.
   C. MOUTH.
   D. STOMACH.

1.	Produces only one enzyme, carbohydrase.
2.	Produces only one enzyme, protease.
3.	Produces no enzymes but reabsorbs excess water.
4.	Produces three enzymes, carbohydrase, protease and lipase.

2. These sentences are about the functions of parts of the blood. Match words from the list with each of the spaces 1-4 in the sentences.
   A. PLATELETS.
   B. WHITE CELLS.
   C. RED CELLS.
   D. PLASMA.
   The watery substance which contains the other blood components is called the____1____. The component which forms part of the body's defence mechanism are the____2____, whereas the____3____ transport oxygen from the lungs to the rest of the body. The____4____ form part of the clotting mechanism of the blood.

3. Different parts of the digestive system produce different enzymes. Choose from the list TWO parts of the digestive system which produce the enzyme protease.
   A. LARGE INTESTINE.
   B. STOMACH.
   C. MOUTH.
   D. LIVER.
   E. PANCREAS.

4. Blood is pumped through the blood vessels by the heart. This is our transport system.

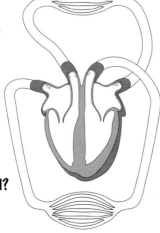

4.1 Which of the following substances is NOT transported by the blood?
   A. HORMONES.
   B. OXYGEN.
   C. URINE.
   D. HEAT.

4.2 Veins ...
   A. Have thick muscular walls which contain elastic fibres.
   B. Only carry oxygenated blood.
   C. Carry blood away from the heart.
   D. Contain valves to prevent backflow.

4.3 The left ventricle is ...
   A. A small chamber carrying oxygenated blood.
   B. A small chamber carrying deoxygenated blood.
   C. A large chamber carrying oxygenated blood.
   D. A large chamber carrying deoxygenated blood.

4.4 The correct pathway of the circulation is ...
   A. Heart ⟶ Capillaries ⟶ Veins ⟶ Arteries.
   B. Heart ⟶ Arteries ⟶ Capillaries ⟶ Veins.
   C. Heart ⟶ Veins ⟶ Capillaries ⟶ Arteries.
   D. Heart ⟶ Arteries ⟶ Veins ⟶ Capillaries.

5. The lungs are designed to enable gases to be efficiently exchanged.

5.1 The smaller branches of the bronchi are called ...
   A. The ALVEOLI.
   B. The TRACHEA.
   C. The BRONCHIOLES.
   D. The AIR PIPES.

5.2 The gases exchanged in the lungs are ...
   A. Carbon Dioxide and Nitrogen.
   B. Carbon Dioxide and Water vapour.
   C. Carbon Dioxide and Oxygen.
   D. Carbon Dioxide, Oxygen and Nitrogen.

5.3 Breathing is necessary because ...
   A. The lungs need to be filled with air.
   B. Carbon Dioxide is needed by body cells to respire.
   C. Oxygen is needed by body cells to respire.
   D. Both Oxygen and Carbon Dioxide are needed by the body cells to respire.

5.4 The Wind Pipe or Trachea splits into ...
A. Lots of Bronchioles.
B. Two Bronchi.
C. Lots of Air Sacs.
D. Lots of Bronchi.

6. Respiration is the process by which cells release energy. It can occur aerobically or anaerobically.

6.1 Aerobic respiration produces ...
A. Carbon Dioxide, Water and Energy.
B. Carbon Dioxide, Glucose and Energy.
C. Carbon Dioxide, Oxygen and Energy.
D. Carbon Dioxide, Oxygen and Glucose.

6.2 The waste product produced during anaerobic respiration is ...
A. CARBON DIOXIDE.
B. UREA.
C. LACTIC ACID.
D. GLUCOSE.

6.3 The energy that is released during respiration is used for ...
A. Combining Oxygen with Carbon Dioxide.
B. Producing lactic acid.
C. Maintaining a steady body temperature.
D. Removing heat from the atmosphere.

6.4 The energy released in respiration comes originally from ...
A. The sun's energy trapped inside glucose molecules.
B. The molecules of oxygen which combine with glucose.
C. Oxygen alone.
D. The blood.

7. The body needs to be able to defend itself against disease in order to remain healthy.

7.1 Which of the following is NOT true of bacteria?
A. They have cytoplasm.
B. They have a cell wall.
C. Can only reproduce inside living cells.
D. They have no distinct nucleus.

7.2 Which of the following is NOT true of viruses?
A. Consist of a protein coat surrounding a few genes.
B. Can only reproduce inside living cells.
C. They are bigger than bacteria.
D. They may produce toxins (poisons).

7.3 Which of the following is NOT an effective defence against invading microbes?
A. The blood clotting mechanism.
B. Sweat glands.
C. Mucus along the lining of the respiratory system.
D. Unbroken skin.

7.4 Which of the following does NOT form part of the white cell's defence against microbes?
A. Ingestion (eating) of microbes.
B. Production of antibodies.
C. Clotting of the blood.
D. Production of antitoxins.

MAINTENANCE OF LIFE

TYPICAL PLANT CELLS

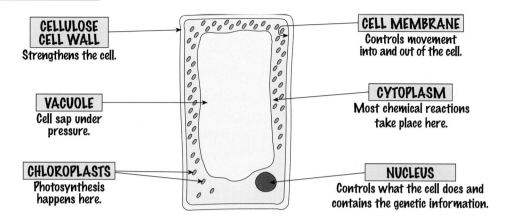

CELLULOSE CELL WALL
Strengthens the cell.

VACUOLE
Cell sap under pressure.

CHLOROPLASTS
Photosynthesis happens here.

CELL MEMBRANE
Controls movement into and out of the cell.

CYTOPLASM
Most chemical reactions take place here.

NUCLEUS
Controls what the cell does and contains the genetic information.

This is a fairly typical plant cell but nevertheless it is SPECIALISED to be particulalrly good at photosynthesis because:

- It has lots of chloroplasts.
- Most of the chloroplasts are near the upper surface to 'catch' the light.
- The length of the cell means that by the time light has reached the bottom of the cell it will most probably have hit a chloroplast.

OTHER SPECIALISED PLANT CELLS

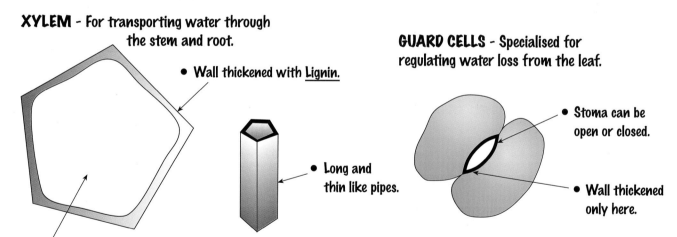

XYLEM - For transporting water through the stem and root.

- Wall thickened with <u>Lignin</u>.

- Long and thin like pipes.

- Empty space - no cytoplasm.

GUARD CELLS - Specialised for regulating water loss from the leaf.

- Stoma can be open or closed.

- Wall thickened only here.

ROOT HAIR CELL - Thin hair-like projections give a big surface area for efficient adsorption.

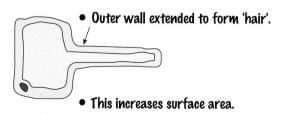

- Outer wall extended to form 'hair'.

- This increases surface area.

A group of cells with similar structure performing the same function is called a TISSUE e.g. XYLEM TISSUE, PALISADE TISSUE. A number of tissues put together form an ORGAN e.g. leaf.

- Key labels: Cellulose Cell Wall, Vacuole, Chloroplasts, Cell Membrane, Cytoplasm and Nucleus.
- Specialised cells include: Xylem, Guard Cells, Root Hair Cells. They all have special features.

GENERAL PLANT STRUCTURE AND ...

STEM - holds plant upright and transports substances between different parts. The stem (or shoot) grows towards light and against the force of gravity.

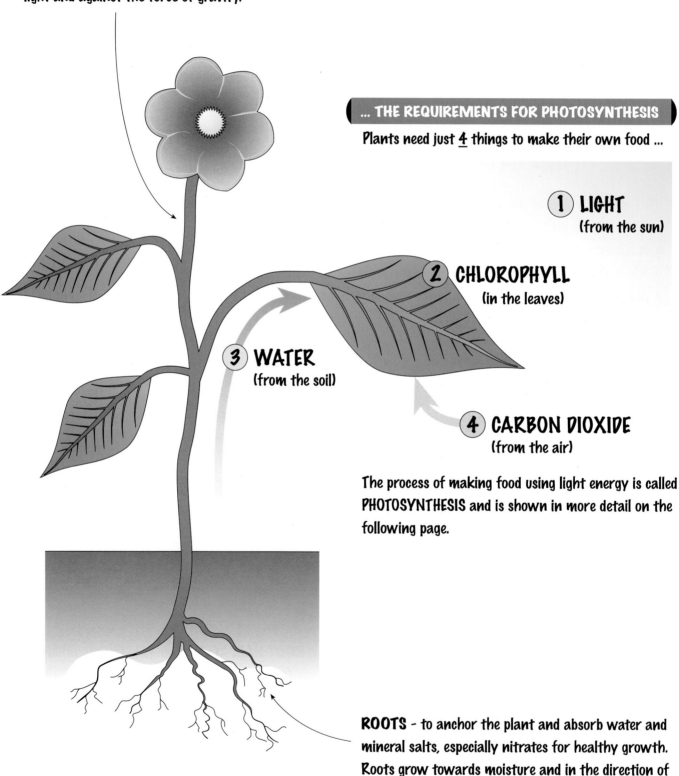

... THE REQUIREMENTS FOR PHOTOSYNTHESIS

Plants need just <u>4</u> things to make their own food ...

1 LIGHT (from the sun)

2 CHLOROPHYLL (in the leaves)

3 WATER (from the soil)

4 CARBON DIOXIDE (from the air)

The process of making food using light energy is called PHOTOSYNTHESIS and is shown in more detail on the following page.

ROOTS - to anchor the plant and absorb water and mineral salts, especially nitrates for healthy growth. Roots grow towards moisture and in the direction of the force of gravity.

- The requirements for photosynthesis are:
  Light (from the Sun), Chlorophyll (in the leaves), Water (from the soil) and Carbon Dioxide (from the air).

PHOTOSYNTHESIS – The Equation

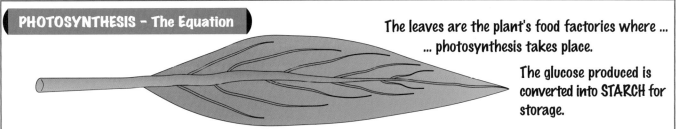

The leaves are the plant's food factories where ...
... photosynthesis takes place.

The glucose produced is converted into STARCH for storage.

$$CARBON\ DIOXIDE + WATER + LIGHT\ ENERGY \Longrightarrow GLUCOSE + OXYGEN$$

CARBON DIOXIDE diffuses into the leaf through the stomata from the atmosphere (See P.6).

WATER passes from the roots through xylem vessels (See P.20) to the leaves.

LIGHT ENERGY is absorbed by the chlorophyll molecules inside the chloroplasts of leaf cells (See P.17).

GLUCOSE is a sugar which may be respired by the cells (See P.10) or stored as insoluble STARCH.

OXYGEN is a by-product of the process and diffuses out of the stomata into the atmosphere (See P.20).

THREE FACTORS WHICH MAY AFFECT THE RATE OF PHOTOSYNTHESIS

The THREE factors which can limit the rate of photosynthesis are ...
- LOW TEMPERATURE   • SHORTAGE OF CARBON DIOXIDE   • SHORTAGE OF LIGHT

LOW TEMPERATURE
- Chemical reactions (e.g. photosynthesis) in plants, like all living things are controlled by **ENZYMES**.
- These are destroyed at temperatures of around **45°C**.
- However plants rarely meet this temperature, but they do experience very cold days, and photosynthesis drops off to next to nothing!

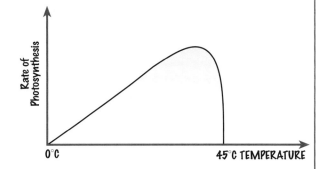

SHORTAGE OF CARBON DIOXIDE
- In the first part of the graph, carbon dioxide is clearly the limiting factor ...
  ... as increasing the concentration increases the rate.
- In the 'flat' part of the graph, increase in concentration now has no effect ...
  ... therefore LIGHT INTENSITY <u>or</u> WARMTH must be the LIMITING FACTOR.

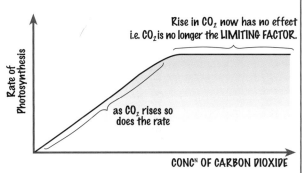

SHORTAGE OF LIGHT
- In the first part of the graph, light is clearly the limiting factor ...
  ... as increasing the light intensity increases the rate.
- In the 'flat' part of the graph, increase in light intensity now has no effect ...
  ... therefore CARBON DIOXIDE CONCENTRATION <u>or</u> WARMTH must be the LIMITING FACTOR.

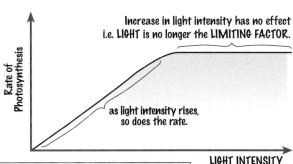

In practice these **THREE** factors interact and any one of them at a particular time may be the factor that limits photosynthesis.

- Photosynthesis equation: Carbon Dioxide + Water + Light Energy $\longrightarrow$ Glucose + Oxygen
- Low temperature, Shortage of Carbon Dioxide or Shortage of Light limit the rate of photosynthesis.

Plants lose water vapour from their leaves in a process called TRANSPIRATION.

THE LEAF

This is a cross-section of a leaf.

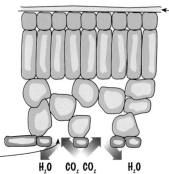

• This **WAXY LAYER** stops too much water from just evaporating away from the leaf.
• Transpiration is more rapid in **HOT, DRY** or **WINDY** conditions, ...

... so plants which live in these conditions have a **THICKER LAYER OF WAX.**

STOMATA —

H_2O CO_2 CO_2 H_2O

RAW MATERIAL No.1 - CARBON DIOXIDE diffuses into the leaf via tiny holes called **STOMATA.**
However taking in carbon dioxide means that the plant must also lose water through the **STOMATA.**

TRANSPIRATION, THEREFORE, IS THE PRICE THE PLANT MUST PAY IN ORDER TO PHOTOSYNTHESISE!!

THE STEM

Flowering plants have **SEPARATE TRANSPORT SYSTEMS** for water and nutrients ...

• **PHLOEM TISSUE ...**

... carries sugars from the leaves to the rest of the plant, especially **GROWING REGIONS** and **STORAGE ORGANS.**

• **XYLEM TISSUE ...**

... transports water and soluble mineral salts from the roots to the stem and leaves. It continually replaces the water lost in transpiration.

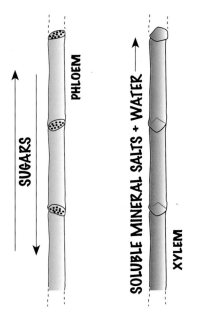

PHLOEM
SUGARS
SOLUBLE MINERAL SALTS + WATER
XYLEM

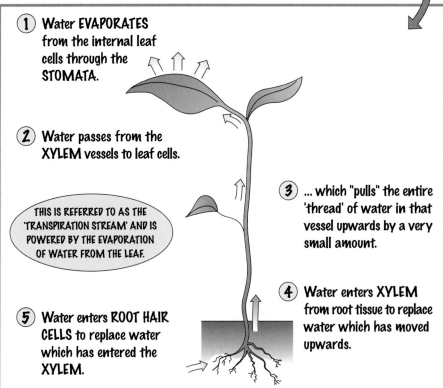

① Water **EVAPORATES** from the internal leaf cells through the **STOMATA.**

② Water passes from the **XYLEM** vessels to leaf cells.

THIS IS REFERRED TO AS THE 'TRANSPIRATION STREAM' AND IS POWERED BY THE EVAPORATION OF WATER FROM THE LEAF.

③ ... which "pulls" the entire 'thread' of water in that vessel upwards by a very small amount.

④ Water enters **XYLEM** from root tissue to replace water which has moved upwards.

⑤ Water enters **ROOT HAIR CELLS** to replace water which has entered the **XYLEM.**

THE ROOT

RAW MATERIAL No.2 - WATER enters the plant via the roots.
Most of it is absorbed by the **ROOT HAIR CELLS.** This water passes into the xylem vessels to replace water which is continually moving up the stem.

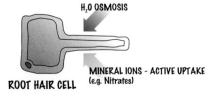

H_2O OSMOSIS

MINERAL IONS - ACTIVE UPTAKE (e.g. Nitrates)

ROOT HAIR CELL

• In transpiration, water passes from the soil into the roots and up the stem to replace water lost from the stomata by evaporation. Evaporation is the driving force of transpiration.

CONTROLLING WATER LOSS

- The size of the stomata is controlled by a pair of **GUARD CELLS**.
- If plants lose water faster than it is replaced by the roots ...

   ... the stomata can be closed to prevent wilting and eventual dehydration.

If the plant has lots of water then the stomata are fully open ...

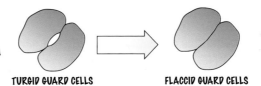

TURGID GUARD CELLS

FLACCID GUARD CELLS

... when water isn't available, then the stomata are closed.

MAINTAINING SUPPORT

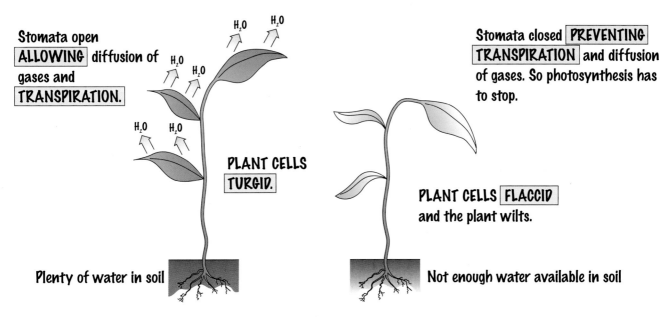

Stomata open $\boxed{\text{ALLOWING}}$ diffusion of gases and $\boxed{\text{TRANSPIRATION.}}$

H_2O H_2O H_2O H_2O H_2O H_2O

PLANT CELLS $\boxed{\text{TURGID.}}$

Plenty of water in soil

Stomata closed $\boxed{\text{PREVENTING}}$ $\boxed{\text{TRANSPIRATION}}$ and diffusion of gases. So photosynthesis has to stop.

PLANT CELLS $\boxed{\text{FLACCID}}$ and the plant wilts.

Not enough water available in soil

When the plant has plenty of water, the cell contents "swell up" and press against the cell wall. This increases the rigidity of the plant tissue and provides the main method of support in young, non-woody stems.

TURGID

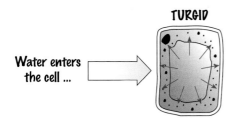

Water enters the cell ...

This causes the cytoplasm and cell contents to press up hard against the cell wall (a bit like a balloon in a shoe box). When this happens in the whole plant, the stem and leaves appear firm and upright.

FLACCID

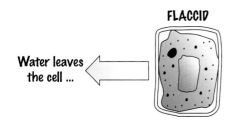

Water leaves the cell ...

When there isn't enough water, the cell contents look almost "shrivelled up" and don't press against the cell wall. This causes the leaves and stem to look limp and droopy.

- Stomata close to prevent water loss (and therefore transpiration) when there is insufficient soil water. Photosynthesis therefore stops.
- Plant cells become turgid when water enters the cell and flaccid when water leaves the cell.

PLANT RESPONSES

Plants are sensitive to: • **LIGHT**   • **MOISTURE**   • **GRAVITY**
* SHOOTS grow TOWARDS LIGHT and AGAINST THE FORCE OF GRAVITY.
* ROOTS grow TOWARDS MOISTURE and in the DIRECTION OF GRAVITY.

These responses are controlled by HORMONES (See P.24) which coordinate and control growth.

Hormones are produced in the growing tips of shoots and roots but can then collect unevenly ...
... causing unequal growth rates in different parts of the plant.

GRAVITY

GERMINATING SEEDLING

In the Shoot ...
* ... Hormone collects on the lower side ...
* ... and stimulates the growth of the cells on this side.
* Therefore shoot grows upwards ...
* ... away from the force of gravity.

In the Root ...
* ... Hormone also collects on the lower side ...
* ... but slows down the growth of the cells on this side.
* Therefore root grows downwards ...
* ... towards the force of gravity.

LIGHT

* In shoots, LIGHT causes HORMONES ...
    * ... to accumulate on the shaded part of the stem ...
        * ... which causes growth on that side ...
            * ... and the plant grows towards the sun.

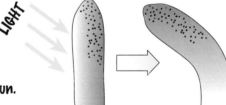

ARTIFICIAL USE OF HORMONES

Farmers do this to increase their yield and to organise ripening times to suit their own convenience.
It's quite unnatural really isn't it?

* ROOTING COMPOUND - Consists of a hormone which encourages ...
                ... the GROWTH OF ROOTS ...
                    ... in STEM CUTTINGS ...
                        ... so lots of plants can be obtained from only one.

* RIPENING HORMONE - Causes plants to ripen at set time ...
                ... sometimes during transport.
                    Is achieved by spraying.

* SELECTIVE WEEDKILLERS - Disrupt the normal growth patterns ...
                ... of their target plants ...
                    ... leaving other plants untouched.

* Shoots grow towards light and against the force of gravity.
* Roots grow towards moisture and in the direction of gravity.
* These responses are controlled by hormones.

The nervous system consists of the BRAIN, the SPINAL CORD, the PAIRED NERVES and RECEPTORS.

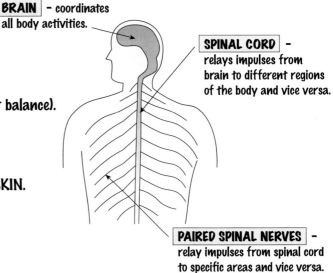

BRAIN – coordinates all body activities.

SPINAL CORD - relays impulses from brain to different regions of the body and vice versa.

PAIRED SPINAL NERVES - relay impulses from spinal cord to specific areas and vice versa.

RECEPTORS

- **LIGHT** RECEPTORS IN THE EYES.
- **SOUND** RECEPTORS IN THE EARS.
- **CHANGES OF POSITION** RECEPTORS IN THE EARS (for balance).
- **TASTE** RECEPTORS ON THE TONGUE.
- **SMELL** RECEPTORS IN THE NOSE.
- **PRESSURE AND TEMPERATURE** RECEPTORS IN THE SKIN.

Information from <u>Receptors</u> passes along <u>Nerves</u> to the <u>Brain</u> which coordinates the response.

RECEPTORS $\implies$ NERVES $\implies$ BRAIN

- Nerve cells are also called **NEURONS** and are specialised cells which conduct **NERVE IMPULSES.**

<u>Receptors</u> respond to **STIMULI** (changes in the environment).

THE STRUCTURE OF THE EYE

The Eye is quite a complicated sense organ which focusses light onto light-sensitive receptor cells in the retina. These are then stimulated and cause nerve impulses to pass along sensory neurons to the brain.

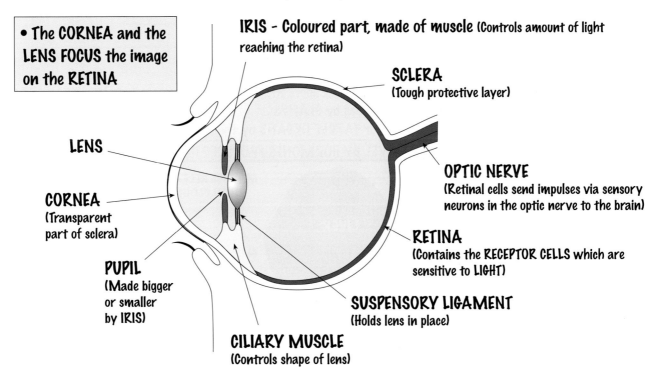

- The CORNEA and the LENS FOCUS the image on the RETINA

IRIS - Coloured part, made of muscle (Controls amount of light reaching the retina)

SCLERA (Tough protective layer)

LENS

OPTIC NERVE (Retinal cells send impulses via sensory neurons in the optic nerve to the brain)

CORNEA (Transparent part of sclera)

RETINA (Contains the RECEPTOR CELLS which are sensitive to LIGHT)

PUPIL (Made bigger or smaller by IRIS)

SUSPENSORY LIGAMENT (Holds lens in place)

CILIARY MUSCLE (Controls shape of lens)

- Receptors respond to various stimuli and pass information to the brain via nerves and the spinal cord.
- The eye is our most complicated receptor. It uses the Cornea and Lens to focus light onto the Retina.

Waste products can't just be left to accumulate inside the body because they would change our INTERNAL ENVIRONMENT. Similarly, other levels have to be controlled so that conditions remain within fairly narrow tolerances. The scientific name for all this is HOMEOSTASIS.

WASTE PRODUCTS WHICH HAVE TO BE REMOVED

CARBON DIOXIDE	• Produced by RESPIRATION. Removed via the LUNGS when we breathe out.
UREA	• Produced by LIVER breaking down excess amino acids. Removed by KIDNEYS, and transferred to bladder before being released.

INTERNAL CONDITIONS WHICH HAVE TO BE CONTROLLED

WATER CONTENT	Water lost by	• breathing via lungs • sweating • excess via kidneys in urine
	Water gained by	• drinking
ION CONTENT (Sodium, Potassium etc.)	Ions are lost by	• sweating • excess via kidneys in urine
	Ions are gained by	• eating • drinking
TEMPERATURE (Ideally at 37°C) - because this is the temperature at which ENZYMES work best!	Temperature increased by:-	• shivering • 'shutting down' skin capillaries
	Temperature decreased by:-	• sweating • 'opening up' skin capillaries
BLOOD GLUCOSE	Blood glucose increased by:-	• Hormone GLUCAGON (from the PANCREAS)
	Blood glucose decreased by:-	• Hormone INSULIN (from the PANCREAS)

HORMONES

Many processes within the body (including control of some of the above internal conditions) are coordinated by HORMONES.
These are ...   • CHEMICAL 'MESSENGERS', produced by GLANDS ...
                • ... which are transported to their TARGET ORGANS by the BLOODSTREAM.

EXAMPLE-THE CONTROL OF BLOOD SUGAR LEVEL BY HORMONES PRODUCED IN THE PANCREAS

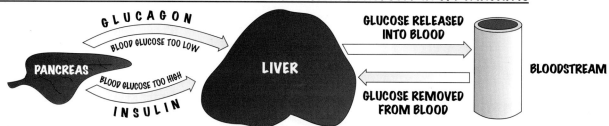

After a meal, the blood level of glucose rises as a result of carbohydrate digestion. The liver then starts to store this glucose and releases it only when the blood sugar level starts to fall.

DIABETES

• Is a condition where the PANCREAS does not produce enough INSULIN, causing patient's BLOOD GLUCOSE LEVEL to RISE to possibly fatally high levels.
• The condition can be controlled by careful diet or injection of INSULIN.

F
o
• Homeostasis is the maintenance of a constant internal environment.
• Blood sugar level is controlled by hormones produced in the pancreas.

DRUGS are chemicals which can affect the behaviour of humans. They may be obtained from LIVING THINGS, or may be SYNTHETIC (MAN MADE). In this section of the syllabus, we concentrate on three broad types:

- SOLVENTS
- TOBACCO
- ALCOHOL

SOLVENTS

Solvents are chemicals which dissolve substances which are not soluble in water.

They are used in:
- GLUES
- POLISHES
- PAINTS
- HAIRSPRAYS
- BUTANE GAS
- DRY CLEANING FLUID

The vapour from these various household substances is inhaled. This practice can cause the following problems ...

- **HALLUCINATIONS** — Users may lose their grip on reality.
- **PERSONALITY CHANGE** — Users may start to display different personality traits.
- **DAMAGE TO ORGANS** — Including Lungs, Brain, Liver and Kidneys. This is usually permanent.

TOBACCO

Tobacco contains NICOTINE, and when burnt and inhaled produces TAR and CARBON MONOXIDE. Tobacco is a legal substance which causes lots of health problems ...

- **LUNG CANCER** — caused by substances in the tar of cigarette smoke.
- **LUNG INFECTIONS** — due to cilia (See P.6) not working because of the tar.
- **EMPHYSEMA** — air sacs damaged through coughing.
- **BRONCHITIS** — irritation caused by smoke particles, and increased mucus ...

  ... allows infection to occur more easily.
- **ARTERIAL DISEASE** — leading to heart attacks, strokes and even amputations, due to narrowing of

  blood vessels.

ALCOHOL

In moderation this is relatively harmless. However alcohol abuse and dependency can cause serious problems ...

- **LIVER DAMAGE** — Alcohol is a mild poison and causes parts of the liver ...

  ... to become fibrous and therefore useless.
- **BRAIN DAMAGE** — Regular doses lead to increased brain cell death ...

  ... and a drop in mental performance e.g. memory.
- **IMPAIRED JUDGEMENT** — This is while under the influence of alcohol. Lack of self-control can result in ...

  ... acts of bravado or foolhardiness which may have fatal results.
- **SLOWS REACTIONS** — Again while under the influence, the whole nervous system is slowed down,

  resulting in unconsciousness or even coma.
- **ADDICTION** — Extreme alcohol dependency can lead to regular days off, ...

  ... reduced performance at work, violence and money problems.

- Drugs are chemicals which can affect the behaviour of humans.
- They can be Natural or Synthetic but to a greater or lesser extent they are always harmful in excess.

1. Choose from the list to match the numbers 1-4 below.
   A. CYTOPLASM.
   B. CHLOROPLAST.
   C. NUCLEUS.
   D. CELL MEMBRANE.

1.	Photosynthesis occurs in this structure.
2.	Controls what the cell does and contains the genetic information.
3.	Most chemical reactions take place in this.
4.	Controls movement into and out of the cell.

2. These sentences are about photosynthesis in green plants. Match words from the list with each of the spaces 1-4 in the sentences.
   A. STOMATA.
   B. XYLEM VESSELS.
   C. LIGHT ENERGY.
   D. GLUCOSE.

   Carbon dioxide diffuses into the leaf through the
   _____1_____. The chlorophyll molecules inside the
   chloroplasts absorb_____2_____. Water passes from
   the roots through the _____3_____ to the leaves.
   _____4_____is a sugar which is produced during
   photosynthesis.

3. The body has to control certain internal conditions. Which **TWO** of the following are NOT controlled to some extent by the kidney?
   A. WATER CONTENT.
   B. ION CONTENT.
   C. TEMPERATURE LEVEL.
   D. UREA CONTENT.
   E. BLOOD GLUCOSE LEVEL.

4.
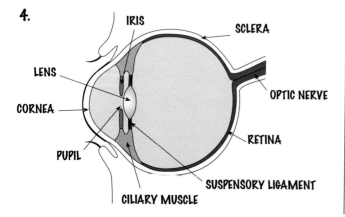

The diagram of the eye shows some of the important structures in it.

4.1 Which one of these structures controls the amount of light entering the eye?
   A. CORNEA.
   B. LENS.
   C. IRIS.
   D. CILIARY MUSCLE.

4.2 Which one of these structures contain the light-sensitive cells which send impulses to the brain?
   A. OPTIC NERVE.
   B. RETINA.
   C. SCLERA.
   D. CORNEA.

4.3 Which of the following are not receptors?
   A. The fine hairs covering the arms and legs.
   B. The parts of the ears which enable us to keep our balance.
   C. The cells in the skin which detect pressure and temperature changes.
   D. The cells on the tongue which can detect different chemicals.

4.4 Which TWO structures in the eye actually focus the image on the retina?
   A. LENS AND CORNEA.
   B. LENS AND PUPIL.
   C. CORNEA AND PUPIL.
   D. CORNEA AND CILIARY MUSCLE.

5. Plants need to obtain raw materials and move substances to different regions within the plant.

5.1 How do plants obtain water?
   A. It diffuses into stem cells.
   B. It is absorbed into root hair cells.
   C. It passes into stomata in root hair cells.
   D. It diffuses into stomata in leaves.

5.2 How is water transported to the growing regions of the plant?
   A. From cell to cell.
   B. Through phloem tissue.

C. Through xylem tissue.

D. Through stomata.

5.3 How are the products of photosynthesis transported through the plant?

A. From cell to cell.

B. Through phloem tissue.

C. Through xylem tissue.

D. By diffusion through the stem.

5.4 How does carbon dioxide get to the sites of photosynthesis?

A. Through the stomata and then to the chloroplasts.

B. By diffusion through the stomata and then to the chloroplasts.

C. By diffusion through the leaf's surface and then to the chloroplasts.

D. By diffusion through the plant stem and then to the chloroplasts.

6. The diagram opposite shows a young plant developing from a seed.

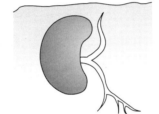

6.1 The root grows out of the seed and ...

A. Away from the light.

B. Away from the force of gravity.

C. Towards water.

D. Towards the force of gravity.

6.2 The stem (or shoot) grows out of the seed and ...

A. Towards the light.

B. Away from the force of gravity.

C. Away from water.

D. Towards the air.

6.3 The stem (or shoot) turns upwards due to ...

A. An equal distribution of plant hormones.

B. An equal distribution of water.

C. An unequal distribution of hormones.

D. An unequal distribution of chloroplasts in the stem.

6.4 Which of the following are plants **NOT** sensitive to?

A. Light.

B. Air pressure.

C. Moisture.

D. Gravity.

7. The body needs to maintain blood sugar level between fairly narrow limits.

7.1 The organ responsible for controlling blood sugar is ...

A. The small intestine.

B. The liver.

C. The pituitary gland.

D. The pancreas.

7.2 Urea is released in the urine and is produced by ...

A. The liver breaking down excess amino acids.

B. The liver breaking down glucose.

C. The pancreas breaking down excess amino acids.

D. The pancreas breaking down glucose.

7.3 The water content of the body is important. What is lost via ...

A. The liver, the lungs, and the kidneys.

B. The liver, the stomach and the kidneys.

C. The skin, the lungs and the kidneys.

D. The skin, the lungs and the stomach.

7.4 Diabetes is a disease in which a person ...

A. Cannot produce enough insulin.

B. Cannot produce enough glucagon.

C. Cannot produce enough glycogen.

D. Cannot produce enough glucose.

METALS

METALS AND NON-METALS IN THE PERIODIC TABLE

As we have seen, the PERIODIC TABLE ...
... arranges ALL THE ELEMENTS ...
... in order of INCREASING PROTON NUMBER.
It's just a way of CLASSIFYING the elements.

> MORE THAN THREE QUARTERS OF THE ELEMENTS ARE METALS.

> LESS THAN ONE QUARTER OF THE ELEMENTS ARE NON-METALS.

NON-METALS

METALS

COMPARING METALS AND NON-METALS

METALS	NON - METALS
• All are SOLIDS at room temperature (except Mercury which is a liquid.)	• Half of them are GASES, and Bromine is a LIQUID at room temperature.
• Have HIGH MELTING POINTS	• Have LOW MELTING POINTS and BOILING POINTS
• Are SHINY at least when freshly cut	• Are mostly DULL
• Can be HAMMERED and BENT into shape. Usually, tough and strong.	• Usually BRITTLE and CRUMBLE easily when solid.
• GOOD CONDUCTORS of heat and electricity when solid or liquid.	• POOR CONDUCTORS of heat and electricity when solid or liquid.
• Form ALLOYS (mixtures of metals)	• Don't (obviously!) form alloys.

HOW THESE PROPERTIES ARE PUT TO GOOD USE IN THREE COMMON METALS

COPPER, Cu

• Is mixed with tin to form BRONZE ...
... and zinc to form BRASS.

CAN FORM ALLOYS

• Is used for PIPES ...
... in PLUMBING.

CAN BE HAMMERED AND BENT INTO SHAPE

• Is used for ELECTRICAL WIRING.

GOOD ELECTRICAL CONDUCTOR

IRON, Fe

• Is used for CAST IRON SAUCEPANS.

GOOD HEAT CONDUCTOR

• Is mixed with CARBON + ...
... small quantities of other metals ...
... to make STEEL.

CAN FORM ALLOYS

• Is used for CAR ENGINE BLOCKS.

STRONG AND CONDUCTS HEAT AWAY WELL

ALUMINIUM, Al

• Because it has a LOW DENSITY ...
... it is used for AIRCRAFT BODYWORK.

STRONG AND CAN BE BENT

• Because it becomes covered ...
... in a layer of aluminium oxide, ...
... it does not need to be painted.
• It is used therefore for ...
... GREENHOUSE and WINDOW FRAMES.

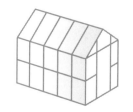

STRONG AND CAN BE BENT

• More than three quarters of the elements in the periodic table are metals, the others are non-metals.
• Different metals have different uses because of the properties they have.

By observing how metals react with OXYGEN (air), WATER, and DILUTE ACID, we can place them in order of how REACTIVE they are. This is called the REACTIVITY SERIES.

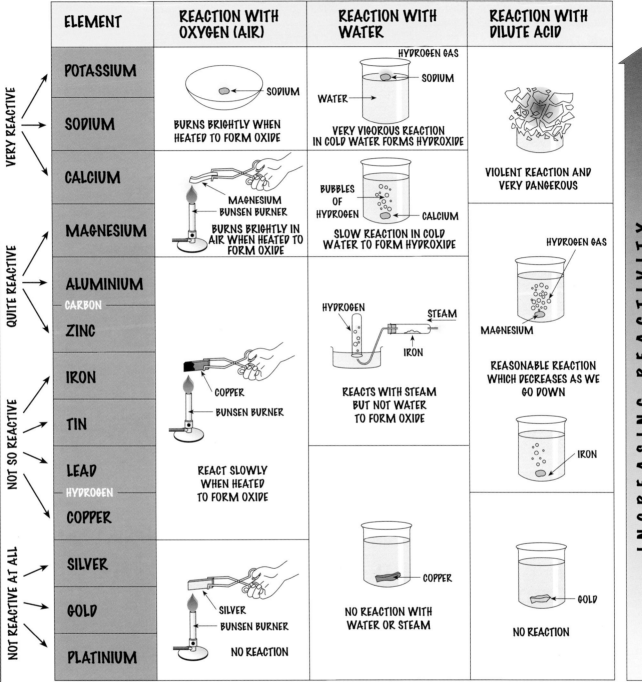

ELEMENT	REACTION WITH OXYGEN (AIR)	REACTION WITH WATER	REACTION WITH DILUTE ACID
VERY REACTIVE POTASSIUM, SODIUM, CALCIUM	BURNS BRIGHTLY WHEN HEATED TO FORM OXIDE (SODIUM)	HYDROGEN GAS — SODIUM — WATER. VERY VIGOROUS REACTION IN COLD WATER FORMS HYDROXIDE	VIOLENT REACTION AND VERY DANGEROUS
QUITE REACTIVE MAGNESIUM, ALUMINIUM, — CARBON —, ZINC	MAGNESIUM — BUNSEN BURNER — BURNS BRIGHTLY IN AIR WHEN HEATED TO FORM OXIDE	BUBBLES OF HYDROGEN — CALCIUM. SLOW REACTION IN COLD WATER TO FORM HYDROXIDE	HYDROGEN GAS — MAGNESIUM. REASONABLE REACTION WHICH DECREASES AS WE GO DOWN
NOT SO REACTIVE IRON, TIN, LEAD, — HYDROGEN —, COPPER	COPPER — BUNSEN BURNER — REACT SLOWLY WHEN HEATED TO FORM OXIDE	HYDROGEN — STEAM — IRON. REACTS WITH STEAM BUT NOT WATER TO FORM OXIDE	IRON
NOT REACTIVE AT ALL SILVER, GOLD, PLATINIUM	SILVER — BUNSEN BURNER — NO REACTION	COPPER. NO REACTION WITH WATER OR STEAM	GOLD. NO REACTION

INCREASING REACTIVITY

REACTION WITH AIR Nearly all the metals react with OXYGEN to form OXIDES.

METAL + OXYGEN ⟶ METAL OXIDES.

REACTION WITH WATER Some metals react to produce HYDROXIDES (or OXIDES) and HYDROGEN.

METAL + WATER ⟶ METAL HYDROXIDE (or OXIDE) + HYDROGEN

REACTION WITH DILUTE ACIDS Many metals react with acids ...

... to produce a SALT and HYDROGEN.

METAL + ACID ⟶ METAL SALT + HYDROGEN

- The Hydrogen given off in these reactions can be identified in the following way ... TEST FOR HYDROGEN

HYDROGEN LIGHTED SPLINT POP!!!

- The reactivity series lists metals in order of their reactivity based on their reactions with oxygen (air), water and dilute acid.

A DISPLACEMENT REACTION is one in which a MORE REACTIVE metal displaces a LESS REACTIVE metal from a compound in a chemical reaction.

There's just one "<u>incredibly important rule</u>" to remember!

> ### IF THE PURE METAL IS HIGHER IN THE REACTIVITY SERIES THAN THE METAL IN THE COMPOUND, THEN DISPLACEMENT WILL HAPPEN.

Let's consider what happens when an IRON NAIL is put into a beaker of COPPER SULPHATE SOLUTION.

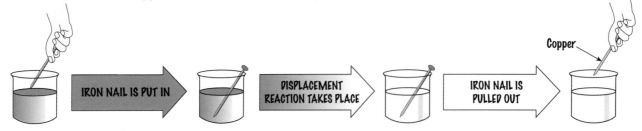

If we look at the POSITIONS of IRON and COPPER in the REACTIVITY SERIES, then what's gone on above can easily be explained

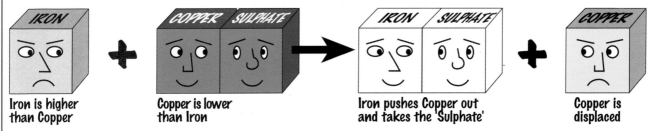

Iron is higher than Copper

Copper is lower than Iron

Iron pushes Copper out and takes the 'Sulphate'

Copper is displaced

This is why an iron nail becomes coated with Copper when it's put in Copper Sulphate Solution, and why the blue solution becomes clear Iron Sulphate solution.

SOME MORE EXAMPLES OF DISPLACEMENT

Example No. 1 ZINC + COPPER SULPHATE SOLUTION
Remember the "Rule". Which is HIGHER in the Reactivity series?

ZINC + COPPER SULPHATE ⟶ ZINC SULPHATE + COPPER

Yes! Zinc is higher so it displaces the copper forming Zinc Sulphate.

Example No. 2 COPPER + LEAD NITRATE SOLUTION
Remember the "Rule". Which is HIGHER in the Reactivity series?

COPPER + LEAD NITRATE ⟶ LEAD NITRATE + COPPER

No! Copper is lower in the series than Lead so no reaction takes place.

SUMMING UP

- Look at the Reactivity Series ...
- ... If the pure metal is higher than the metal in the metal compound ...
- ... then simply swop the metals around!!
- If it isn't there's no reaction.

- If a pure metal is higher in the reactivity series than the metal in the compound then a displacement reaction will happen.
- If the pure metal is lower then no displacement reaction takes place.

The Earth's crust contains many different METALS and METAL COMPOUNDS mixed in with other substances. A metal or metal compound found in enough concentration so that economically viable amounts of the metal can be extracted, is an ORE.

- Most ores are often the OXIDES of metals or substances that can easily be changed into a metal oxide.
- To extract the metal OXYGEN MUST BE REMOVED from the metal oxide.
- This is REDUCTION and the method of EXTRACTION ...
- ... depends on the POSITION of the metal in the REACTIVITY SERIES.

POSITION OF METAL	EXTRACTION PROCESS	EXAMPLES
Metals HIGH in the series (i.e. ABOVE CARBON) ...	... are extracted by ELECTROLYSIS.	ALUMINIUM, Magnesium, Sodium.
Metals in the MIDDLE of the series (i.e. BELOW CARBON) ...	... are extracted by HEATING WITH CARBON.	IRON, Lead, Copper.
Metals VERY LOW in the series ...	... are so UNREACTIVE they exist NATURALLY and NO EXTRACTION IS NEEDED.	GOLD, Platinum.

EXTRACTION OF IRON – The Blast Furnace

it is called a blast furnace because they blast Hot air in.

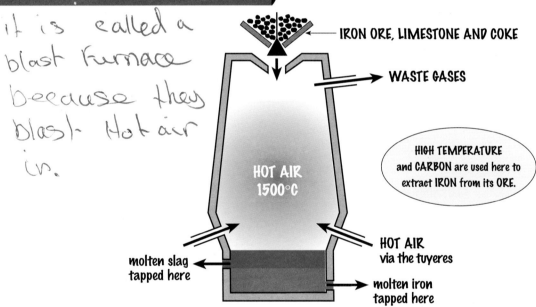

IRON ORE, LIMESTONE AND COKE

WASTE GASES

HIGH TEMPERATURE and CARBON are used here to extract IRON from its ORE.

HOT AIR 1500°C

HOT AIR via the tuyeres

molten slag tapped here

molten iron tapped here

- IRON ORE, LIMESTONE and COKE (CARBON) are fed into the top of the furnace.
- HOT AIR is BLASTED through the furnace via the TUYERES
- OXYGEN from this air reacts with the coke to form CARBON DIOXIDE
- CARBON DIOXIDE reacts with more coke to form CARBON MONOXIDE.

Now at these temperatures CARBON MONOXIDE will very quickly change to CARBON DIOXIDE by taking OXYGEN from the IRON OXIDE

IRON OXIDE + CARBON MONOXIDE ⟶ IRON + CARBON DIOXIDE

Carbon itself is often used to reduce oxides as it is quite high in the reactivity series. However, here it is CARBON MONOXIDE which acts as the REDUCING AGENT.

- The limestone ends up removing certain impurities (especially sand) by combining with it to form SLAG which is then run off.

- Metals above carbon in the reactivity series are extracted by electrolysis.
- Metals in the middle of the series are extracted by heating with carbon. • Metals very low in the series exist naturally.

Metals high in the REACTIVITY SERIES (i.e. <u>ABOVE CARBON</u>) are extracted by ELECTROLYSIS.

- ELECTROLYSIS is the breaking down of a compound ...
    - ... containing IONS ...
        - ... into its ELEMENTS ...
            - ... by using an ELECTRIC CURRENT.
- IONS are atoms or groups of atoms which have an ELECTRICAL CHARGE (+ or -)

EXTRACTION OF ALUMINIUM BY ELECTROLYSIS

- The RAW MATERIALS are ...
    ... PURIFIED ALUMINIUM OXIDE and CRYOLITE.
- The ALUMINIUM OXIDE is dissolved in <u>MOLTEN</u> CRYOLITE at about 850°C, ...
    ... the temperature would have to be much higher if we used just the oxide!
- The ELECTRODES are made of CARBON (graphite) which are used only to conduct the electricity, ...
    ... they don't take any part in the extraction.

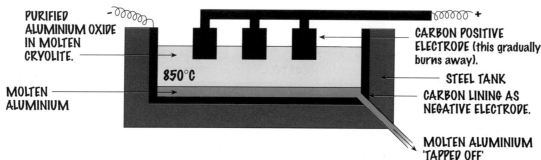

PURIFIED ALUMINIUM OXIDE IN MOLTEN CRYOLITE.

850°C

MOLTEN ALUMINIUM

CARBON POSITIVE ELECTRODE (this gradually burns away).

STEEL TANK
CARBON LINING AS NEGATIVE ELECTRODE.

MOLTEN ALUMINIUM 'TAPPED OFF'

- When a CURRENT passes through the molten mixture ...
- ... AT THE NEGATIVE ELECTRODE ...
- ... POSITIVELY CHARGED ALUMINIUM IONS MOVE TOWARDS IT and ALUMINIUM FORMS and ...
- ... AT THE POSITIVE ELECTRODES ...
- ... NEGATIVELY CHARGED OXYGEN IONS MOVE TOWARDS THEM and OXYGEN FORMS.
- This causes the positive electrodes to burn away quickly and they frequently have to be replaced.

PURIFICATION OF COPPER BY ELECTROLYSIS

Copper can easily be extracted by REDUCTION but when it is needed in a pure form it is purified by ELECTROLYSIS.

- The POSITIVE ELECTRODE is made of IMPURE COPPER.
- AT THE POSITIVE ELECTRODE COPPER IONS pass into the solution.
- AT THE NEGATIVE ELECTRODE COPPER IONS MOVE TOWARDS IT ...
    ... TO FORM COPPER ATOMS ...
    ... which stick to the pure copper electrode.

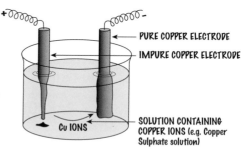

PURE COPPER ELECTRODE
IMPURE COPPER ELECTRODE

Cu IONS

SOLUTION CONTAINING COPPER IONS (e.g. Copper Sulphate solution)

- The impurities fall to the bottom as the IMPURE POSITIVE ELECTRODE GRADUALLY DISSOLVES.

- Electrolysis is the breaking down of a compound containing ions into its elements using an electric current.
- Copper can easily be extracted by reduction but electrolysis is used when the metal is required in its pure form.

INDICATORS

These are really useful dyes which CHANGE COLOUR ...
- ... depending on whether they're in ACIDIC or ALKALINE solutions.
- Some are just simple substances such as LITMUS which changes from RED to BLUE or vice versa ...
- ... whereas others are mixtures of dyes such as UNIVERSAL INDICATOR which ..
- ... show a RANGE OF COLOUR to indicate just how ACIDIC or ALKALINE a substance is.

THE pH SCALE

This measures the ACIDITY, ALKALINITY, or NEUTRALITY of a solution ...

<u>... ACROSS A 14 POINT SCALE</u>

VERY ACIDIC	← →	SLIGHTLY ACIDIC	NEUTRAL	SLIGHTLY ALKALINE	← →	VERY ALKALINE
1 2 3	4 5	6	7	8 9 10	11 12 13	14

When used with UNIVERSAL INDICATOR, we get the following range of colours:-

e.g. Battery Acid / Stomach Acid   Lemon Juice / Vinegar   Soda Water   Water   Soap   Baking Powder   Washing Soda   Oven Cleaner   Potassium Hydroxide

Strong Acid   *Neutral*   *Strong Alkaline*

NEUTRALISATION

Basically, this occurs when the right amounts of acid and alkali react ...

... to "cancel" each other out to form a salt and water (which is NEUTRAL).

ACID	+	ALKALI	⟶	SALT	+	WATER

e.g.   HCl   +   KOH   ⟶   KCl   + H_2O

HYDROCHLORIC ACID   POTASSIUM HYDROXIDE   POTASSIUM CHLORIDE   WATER

STRONG ACID, pH 1   STRONG ALKALI, pH 14   NEUTRAL, pH 7

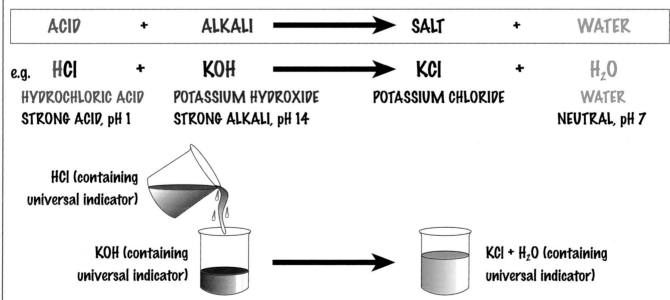

HCl (containing universal indicator)

KOH (containing universal indicator)

KCl + H_2O (containing universal indicator)

(N.B.)

The demonstration above will only work if ...

... BOTH BEAKERS CONTAIN THE SAME NUMBER ...

... OF ACID AND ALKALI MOLECULES, SO THAT THEY NEUTRALISE EACH OTHER EXACTLY.

- Indicators change colour depending on whether they are in acidic or alkaline solutions.
- The pH scale measures the acidity, alkalinity or neutrality of a solution.
- Acid + Alkali ⟶ Salt + Water

COMMON ACIDS AND ALKALIS

As we have seen an **ACID** reacts with an **ALKALI** to produce a **SALT** and WATER. The particular salt produced depends on ...

- ... the METAL in the ALKALI ...
- ... and the ACID USED.

- The three most common acids you will come across are ...

    HYDROCHLORIC ACID   SULPHURIC ACID   NITRIC ACID

- ... and the three most common alkalis you will come across are.

    SODIUM HYDROXIDE   POTASSIUM HYDROXIDE   CALCIUM HYDROXIDE

What we get when we react one the acids above with one of the alkalis can best be summarised in a table.

	HYDROCHLORIC ACID	SULPHURIC ACID	NITRIC ACID
+ SODIUM HYDROXIDE	→ SODIUM CHLORIDE + WATER	→ SODIUM SULPHATE + WATER	→ SODIUM NITRATE + WATER
+ POTASSIUM HYDROXIDE	→ POTASSIUM CHLORIDE + WATER	→ POTASSIUM SULPHATE + WATER	→ POTASSIUM NITRATE + WATER
+ CALCIUM HYDROXIDE	→ CALCIUM CHLORIDE + WATER	→ CALCIUM SULPHATE + WATER	→ CALCIUM NITRATE + WATER

You will have noticed that neutralising ...

- HYDROCHLORIC ACID produces <u>CHLORIDE</u> salts.
- SULPHURIC ACID produces <u>SULPHATE</u> salts.
- NITRIC ACID produces <u>NITRATE</u> salts.

METAL OXIDES AND HYDROXIDES

The ALKALIS mentioned above are made by dissolving the METAL OXIDES or HYDROXIDES in WATER.

SODIUM
POTASSIUM ——→ OXIDES or HYDROXIDES dissolve in water to form ALKALIS.
CALCIUM

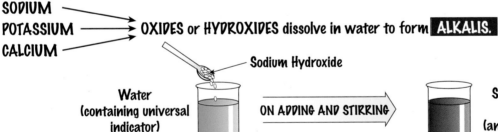

Sodium Hydroxide

Water (containing universal indicator)   ON ADDING AND STIRRING   Sodium Hydroxide solution (an alkaline solution)

NON-METAL OXIDES

In contrast to the above, soluble OXIDES of NON-METALS produce acidic solutions:

CARBON DIOXIDE
SULPHUR DIOXIDE ——→ dissolve in water to form ACIDS.
NITROGEN DIOXIDE

Carbon Dioxide →
Water (containing universal indicator)   AFTER A FEW SECONDS   Carbon Dioxide has dissolved in the water to form an acidic solution

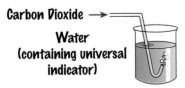

- The particular salt produced when an acid reacts with an alkali depends on the metal in the alkali and the acid used.
- Metal oxides or hydroxides dissolved in water form alkaline solutions.
- Soluble oxides of non-metals dissolve in water producing acidic solutions.

1. Use the words,
   A • METALS or
   B • NON-METALS
   to complete the following sentences.

   Metals are all solids at room temperature apart from Mercury. _Non 2 Metals_ are usually brittle and crumble easily when solid.
   Metals 3 are all good conductors of heat and electricity, whereas _Non 4 Metals_ are poor conductors.

2. <u>Sodium</u> reacts faster than Calcium with cold water. <u>Copper</u> displaces silver from silver nitrate solution. <u>Zinc</u> reacts more slowly than <u>magnesium</u> with dilute acid.
   Match each of the underlined metals with each of the numbers 1 - 4 in the reactivity series.

   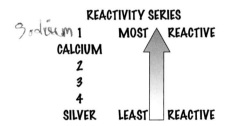

   REACTIVITY SERIES
   Sodium 1   MOST ▲ REACTIVE
   CALCIUM
   2
   3
   4
   SILVER   LEAST REACTIVE

3. Copper is a metal which is used to make electrical wiring. Which <u>TWO</u> properties of copper make it a good choice?
   A • It is a good conductor. ✓
   B • It is a shiny metal.
   C • It is a good conductor of electricity. ✓
   D • It has a high melting point.
   E • It can be bent and pulled into different shapes.

4. The diagram below shows apparatus which can be used to purify copper. Which <u>TWO</u> of the following statements are true regarding this process?

   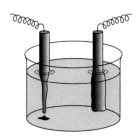

A • The positive electrode is made of pure copper.
B • Copper ions travel from the +ve electrode to the -ve electrode.
C • Copper ions travel from the -ve electrode to the +ve electrode.
D • The solution contains copper ions.
E • The negative electrode is made of impure copper.

5. The following chart shows the colour that universal indicator becomes at different pH values:

Colour:	Red	Orange	Yellow	Green	Blue	Navy Blue	Purple
pH:	0 - 2	3 - 4	5 - 6	7	8 - 9	10 - 12	13 - 14

ACID   NEUTRAL   ALKALINE

5.1 Carbon dioxide is a non-metal oxide which makes green universal indicator turn ...
   A. Red
   B. Purple
   C. Yellow
   D. Blue

5.2 Sodium hydroxide is a metal hydroxide which makes green universal indicator turn ...
   A. Purple
   B. Green
   C. Red
   D. Orange

5.3 If pure water is added to green universal indicator it ...
   A. Turns yellow
   B. Turns blue
   C. Stays green
   D. Turns orange

5.4 If a strong alkali, Potassium Hydroxide is added to an equal volume of equal strength Hydrochloric Acid, a neutralisation reaction occurs. What is formed in this reaction?
   A. A metal hydroxide and hydrogen.
   B. A salt and hydrogen.
   C. A metal oxide and water.
   D. A salt and water.

6. Aluminium is extracted from Aluminium oxide by dissolving it in molten cryolite at 850°C and passing an electric current through it.

6.1 Which of the following describes best what has happened to the Aluminium oxide?
A. It has been neutralised.
B. It has been reduced.
C. It has been oxidised.
D. It has been purified.

6.2 Aluminium is produced at the negative electrode because ...
A. Positively charged aluminium ions are attracted to it.
B. Negatively charged aluminium ions are attracted to it.
C. Aluminium ions have a negative charge.
D. Aluminium is a metal and conducts electricity.

6.3 Why is cryolite used in this reaction?
A. To break the aluminium oxide into ions.
B. To conduct the electric current through the molten mixture.
C. To reduce the aluminium oxide.
D. To reduce the temperature needed to melt the aluminium oxide.

6.4 Aluminium powder will react with Iron Oxide to form Aluminium Oxide and Iron in a displacement reaction. The reason for this is ...
A. Aluminium is less reactive than iron.
B. Aluminium is higher on the reactivity series than iron.
C. Aluminium is more reactive than oxygen.
D. Aluminium is less reactive than oxygen.

7. Iron can be extracted from iron ore using the blast furnace below.

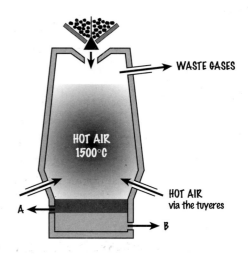

7.1 Which substances are fed into the top of the blast furnace?
A. Iron ore.
B. Iron ore and Coke.
C. Iron ore, Limestone and Coke.
D. Iron ore and Limestone.

7.2 Which two substances marked A + B are tapped off at the bottom of the furnace?
A. Carbon and Iron.
B. Carbon and Slag.
C. Slag and Iron.
D. Carbon, Slag and Iron.

7.3 Which substance is responsible for removing the oxygen from Iron Oxide?
A. Carbon monoxide.
B. Carbon dioxide.
C. Carbon.
D. Oxygen.

7.4 The process by which Iron oxide becomes Iron in the blast furnace is called ...
A. Precipitation.
B. Oxidation.
C. Replacement.
D. Reduction.

EARTH MATERIALS

LIMESTONE

Limestone is a **SEDIMENTARY ROCK** consisting mainly of **CALCIUM CARBONATE**.
It is cheap, easy to obtain and has many uses:

1. NEUTRALISING AGENT

- Excess **ACIDITY** of soils can cause crop failure.
  - Alkalis can be 'washed out' by acid rain.
    - Powdered limestone can correct this ...
      - ... but it works quite slowly.
- However Calcium Carbonate can be heated to produce **CALCIUM OXIDE (QUICKLIME)**.

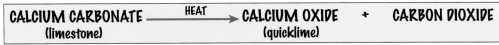

CALCIUM CARBONATE ——HEAT——→ CALCIUM OXIDE + CARBON DIOXIDE
(limestone)   (quicklime)

- This can then be 'SLAKED' with water to produce **CALCIUM HYDROXIDE (SLAKED LIME)**.

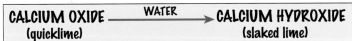

CALCIUM OXIDE ——WATER——→ CALCIUM HYDROXIDE
(quicklime)   (slaked lime)

- This, being a **HYDROXIDE** is quite strongly **ALKALINE** ...
  - ... and so can neutralise soils and lakes ...
    - ... much faster than just using ...
      - ... powdered limestone.

2. BUILDING MATERIAL

- Can be **QUARRIED** and cut ...
  - ... into **BLOCKS** and used directly ...
    - ... to build **WALLS** of houses ...
      - ... in regions where it is plentiful!
        - It is badly affected by **ACID RAIN** ...
          - ... but this takes a long time.

3. GLASS MAKING

- Glass is made by mixing ...
  - ... **LIMESTONE, SAND** and ...
    - ... **SODA** (sodium carbonate) ...
      - ... and heating the mixture until it melts.
        - When cool it is **TRANSPARENT**.

LIMESTONE + SAND + SODA ——HEAT——→ GLASS

4. CEMENT MAKING

- Powdered limestone and powdered **CLAY** ...
  - ... are roasted in a **ROTARY KILN** ...
    - ... to produce dry cement.
      - When the cement is mixed with **WATER, SAND** and **GRAVEL** (crushed rock) ...
        - ... a slow reaction takes place where ...
          - ... a **HARD, STONE-LIKE BUILDING MATERIAL** ...
            - ... called **CONCRETE** is produced.

CEMENT

- Limestone can be used as a neutralising agent, as a building material and for making glass and cement.
- Calcium carbonate ——HEAT——→ Calcium oxide + Carbon dioxide • Calcium oxide ——WATER——→ Calcium hydroxide

The basic facts about the FORMATION, CHARACTERISTICS, and SPECIFIC FEATURES need to be learnt.
We've cut out all the waffle so this is all you need to know.

Rock Type	HOW THEY'RE FORMED	EXAMPLES	SPECIFIC FEATURES OF THESE EXAMPLES	WHAT THEY LOOK LIKE
SEDIMENTARY	• Made from layers of SEDIMENT (small particles) ... • ...whose WEIGHT squeezes out WATER ... • ... causing particles to become "CEMENTED" together ... • ... by SALTS, CRYSTALLISING OUT. YOUNGER ROCKS therefore are usually ON TOP.	SANDSTONE	Formed by particles of SAND, washed down by river, eventually falling to RIVER BED or SEA BED.	• Very GRAINY and CRUMBLY • SANDGRAINS OBVIOUS • Sometimes contains FOSSILS ... • ... which can be used to DATE rocks.
		LIMESTONE	Formed by DEAD REMAINS of SHELLED CREATURES and some INSOLUBLE CALCIUM SALTS. e.g. Calcium Carbonate	• GRAINY + CRUMBLY but less than above • Often contain FOSSILS ... • ... which can be used to DATE rocks.
IGNEOUS	• Formed from MOLTEN ROCK ... • ... called MAGMA ... • ... which wells up from the MANTLE ... • ... and COOLS DOWN, either ... • ... ABOVE (Basalt) or WITHIN (Granite) the earth's crust.	BASALT	Expelled from VOLCANOES. Formed EXTRUSIVELY by cooling, ABOVE the earth's crust.	• Very SMALL CRYSTALS due to FAST COOLING. • Very HARD rocks.
		GRANITE	Magma forced into the earth's crust. Formed INTRUSIVELY by cooling WITHIN the earth's crust.	• LARGE CRYSTALS due to SLOW COOLING • Very HARD rocks.
METAMORPHIC	• Formed by extreme TEMP. and PRESSURE ... • ... caused by MOUNTAIN BUILDING processes ... • ... which force SEDIMENTARY rocks deep underground ... • ... close to MAGMA ... • ... where they become COMPRESSED and HEATED ... • ... changing their TEXTURE and STRUCTURE. • Can be formed from any rock type.	SLATE	Formed when MUDSTONE experiences EXTREME TEMPERATURE and PRESSURE.	• Tiny CRYSTALS form on COOLING • Usually HARD rocks. • Can form BANDS
		MARBLE	Formed when LIMESTONE experiences EXTREME TEMPERATURE and PRESSURE.	• Small CRYSTALS form on COOLING • Usually HARD rocks. • Crystals tend to form BANDS. • BANDING indicates metamorphic rock, eg. Marble, Schist.

- There are three types of rock: Sedimentary, Igneous and Metamorphic.
- Sedimentary rocks are grainy and crumbly and may contain fossils. • Igneous rocks are crystalline and very hard.
- Metamorphic rocks are crystalline and banded.

The ROCK CYCLE is an ongoing CYCLE OF EVENTS where ROCKS at the EARTH'S SURFACE are continually being BROKEN UP, REFORMED and CHANGED.

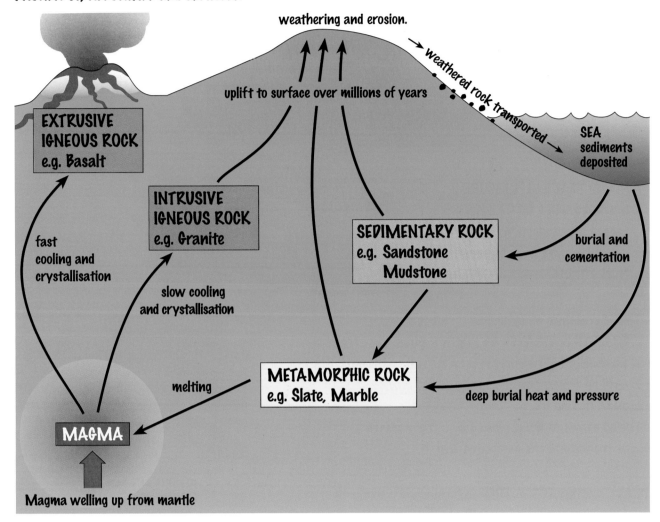

weathering and erosion.

uplift to surface over millions of years

weathered rock transported →

SEA sediments deposited

EXTRUSIVE IGNEOUS ROCK
e.g. Basalt

INTRUSIVE IGNEOUS ROCK
e.g. Granite

SEDIMENTARY ROCK
e.g. Sandstone
Mudstone

fast cooling and crystallisation

slow cooling and crystallisation

burial and cementation

METAMORPHIC ROCK
e.g. Slate, Marble

melting

deep burial heat and pressure

MAGMA

Magma welling up from mantle

ON THE GROUND SURFACE

Remember the 4 stages involved in formation of SEDIMENTARY ROCKS.
- WEATHERING and EROSION.
- TRANSPORTATION of sediment.
- DEPOSITION of sediment.
- BURIAL of sediment.

BELOW THE GROUND SURFACE

1. Molten MAGMA is crystallised to form
   - INTRUSIVE IGNEOUS ROCKS ... ... within the crust.
   - EXTRUSIVE IGNEOUS ROCKS ... ... above the crust.

2. SEDIMENTARY ROCKS get close to the MAGMA and form METAMORPHIC ROCKS.

3. All types of rock can return to the MAGMA by being driven underground very deeply.

- The four stages involved in the formation of sedimentary rocks are: Weathering and Erosion, Transportation, Deposition and Burial. • Molten magma crystallises to form igneous rocks.
  ...morphic rocks are formed when sedimentary rocks get close to the magma.

HOW CRUDE OIL IS FORMED

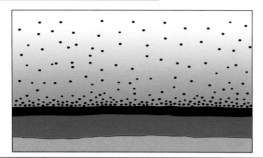

- Formed over millions of years from dead ORGANISMS ...
- ... mainly PLANKTON (tiny sea creatures) ...
- ... which fell to the ocean floor...
- ... and were covered by MUD SEDIMENTS.

- It is a FOSSIL FUEL.

- Action of HEAT and PRESSURE ...
- ... in the ABSENCE OF OXYGEN ...
- ... caused the production of CRUDE OIL ...
- ... which becomes trapped between ...
- ... NON-POROUS layers of sediment.

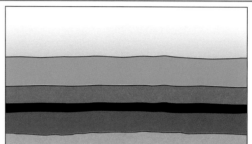

- The oil remained underground (in most cases), ...
- ... TRAPPED in a LAYER OF POROUS ROCK ...
- ... sandwiched between TWO LAYERS OF NON-POROUS ROCK ...
- ... until oil exploration companies drilled down ...
- ... and 'released it'.
- The oil comes to the surface due to the pressure ...
- ... of the natural gas associated with it.

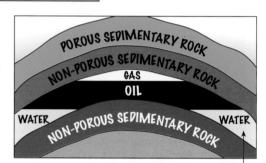

POROUS SEDIMENTARY ROCK

NON-POROUS SEDIMENTARY ROCK

GAS

OIL

WATER NON-POROUS SEDIMENTARY ROCK WATER

POROUS ROCK

OIL AND GAS ARE LESS DENSE THAN WATER ... AND SO THEY RISE TO THE TOP OF THE POROUS ROCK LAYER

COAL , another FOSSIL FUEL, is formed very similarly ...

- ... from the remains of dead PLANTS ...
- ... which were covered by MUD SEDIMENTS etc.

WHAT CRUDE OIL IS

- Crude oil is a mixture of compounds most of which ...
- ... are MOLECULES made up of CARBON and HYDROGEN atoms only, called HYDROCARBONS.

These hydrocarbon molecules vary in size. This affects their properties ...

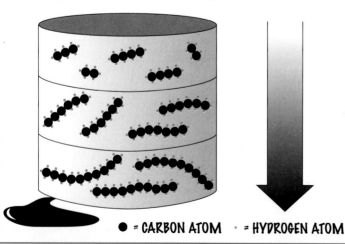

The LARGER the HYDROCARBON:
- the LESS EASILY it FLOWS ...
  ... i.e. the more viscous it is.
- the LESS EASILY it IGNITES ...
  ... i.e. the less flammable it is.
- the LESS VOLATILE it is ...
  ... i.e. it doesn't vaporise as easily.
- the HIGHER IT'S BOILING POINT.

● = CARBON ATOM   · = HYDROGEN ATOM

- Crude oil (a fossil fuel) is mainly made of molecules containing carbon and hydrogen atoms only called hydro

FRACTIONAL DISTILLATION OF CRUDE OIL

Crude oil on it's own isn't a great deal of use. We need to separate it into it's different FRACTIONS all of which have their own PARTICULAR CHARACTERISTICS.

It is possible to separate crude oil into its FRACTIONS by ...

- ... EVAPORATING the oil by heating ...
- ... and then allowing it to CONDENSE ...
- ... at a RANGE of DIFFERENT TEMPERATURES ...
- ... when it will form FRACTIONS each of which ...
- ... will contain molecules ...
- ... with a SIMILAR NUMBER OF CARBON ATOMS.

- This is called FRACTIONAL DISTILLATION ...
- ... and is done in a FRACTIONATING COLUMN.

No. of CARBON ATOMS

COLD

PETROLEUM GAS

70°C PETROL

8

140°C NAPTHA

10

180°C PARAFFIN

14-16

260°C DIESEL OIL

18-22

CRUDE OIL VAPOUR

340°C FUEL OIL

30-40

HOT

40+

BITUMEN

CHEMI

As you move down the fractionating column, the numbers of carbon atoms in the hydrocarbon molecules of each fraction increases.

The greater the number of carbon atom in each hydrocarbon ...

- ... the HIGHER it's BOILING POINT ...
- ... the LESS VOLATILE it is ...
- ... the LESS EASILY IT FLOWS ...
- ... the LESS EASILY IT IGNITES ...

CRACKING

Because the SHORTER CHAIN HYDROCARBONS release energy more quickly by BURNING, there is a greater demand for them.

- Therefore LONGER CHAIN HYDROCARBONS are 'CRACKED' or broken down ...

- ... to produce SHORTER CHAIN HYDROCARBONS.
- Some of these are used as FUELS ...
- ... and some to make PLASTICS such as ...
- ... POLY(ETHENE) ...   ... and PVC.

Plastic Bowls   **Plastic Bags**   **Wellington Boots**   **Electrical Insulation**

- Crude oil is separated into its fractions by fractional distillation and cracking is used to break down long chain hydrocarbons into short chain hydrocarbons. • Short chain hydrocarbons are used as fuels and to make plastics.

- When a substance burns it reacts with OXYGEN ...
- ... to produce compounds called OXIDES.
- When a FOSSIL FUEL burns ...
- ... WASTE PRODUCTS are formed which are released into the ATMOSPHERE.

CARBON DIOXIDE   WATER VAPOUR

- Because these fuels contain CARBON, HYDROGEN and SULPHUR ...
- ... the waste products include CARBON DIOXIDE, WATER VAPOUR (an oxide of hydrogen) ...
- ... and SULPHUR DIOXIDE.

Methane → ← Oxygen

EXAMPLE

NATURAL GAS is made up mainly of METHANE ...

METHANE   +   OXYGEN ———→ CARBON DIOXIDE   +   WATER VAPOUR   +   HEAT

... and if SULPHUR is present in the fuel ...

SULPHUR   +   OXYGEN ———→ SULPHUR DIOXIDE

GLOBAL WARMING - Effects of increasing Carbon Dioxide

- CARBON DIOXIDE is one of the "GREENHOUSE GASES."
- The INCREASED BURNING OF FUELS ...
- ... is INCREASING the LEVELS of CARBON DIOXIDE ...
- ... in the ATMOSPHERE which is resulting in GLOBAL WARMING.

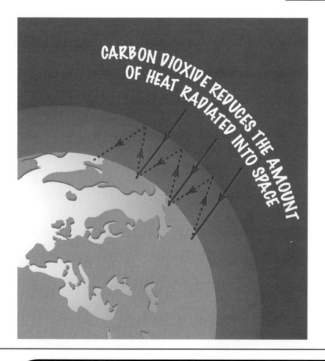

CARBON DIOXIDE REDUCES THE AMOUNT OF HEAT RADIATED INTO SPACE

- Light from the Sun reaches the Earth ...
- ... and passes through the atmosphere.
- This WARMS up the planet which ...
- ... then radiates this heat energy back into SPACE.

- CARBON DIOXIDE helps to trap some of this energy ...
- ... which helps to keep the planet WARM.
- Too much carbon dioxide however leads to ...
- ... too much heat being retained.
- This is GLOBAL WARMING.

- Methane + Oxygen ———→ Carbon dioxide + Water Vapour + Heat
- Sulphur + Oxygen ———→ Sulphur dioxide
- Increased burning of fuels is increasing the level of carbon dioxide in the atmosphere resulting in global warming.

ACID RAIN – Effects of Sulphur Dioxide and Nitrogen Oxides

- SULPHUR DIOXIDE and NITROGEN OXIDES are produced ...
- ... when FUELS are burned in FURNACES and ENGINES (mainly cars).
- These gases then react with WATER VAPOUR in the ATMOSPHERE ...
- ... to produce ACIDS.
- These fall as ACID RAIN.

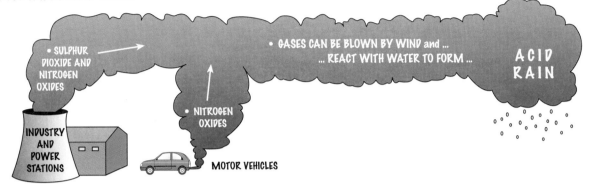

- SULPHUR DIOXIDE AND NITROGEN OXIDES
- GASES CAN BE BLOWN BY WIND and ...
  ... REACT WITH WATER TO FORM ...
  A C I D R A I N
- NITROGEN OXIDES

INDUSTRY AND POWER STATIONS

MOTOR VEHICLES

The GASES themselves can harm PLANTS and ANIMALS directly ...	... while ACID RAIN causes LAKES and RIVERS ... ... to become so ACIDIC that ... ... PLANTS and ANIMALS cannot survive!! ...

... and cause EROSION damage to ... ... STONE and METALWORK of BUILDINGS ...	... it also causes EROSION damage to ... ... STONE and METALWORK of BUILDINGS.

COMPOSITION OF THE ATMOSPHERE

Our atmosphere has been more or less the same for 200 million years!!!

CARBON DIOXIDE, CO_2 (0.03%)

MAINLY ARGON, + OTHER NOBLE GASES (1%)

NITROGEN, N_2 (78%)

OXYGEN, O_2 (21%)

- WATER VAPOUR may also be present in varying quantities (0 - 3%).

- Sulphur dioxide and nitrogen oxides react with water vapour in the atmosphere to produce Acid Rain.
- The atmosphere is made from Nitrogen, Oxygen, Carbon dioxide, Argon (and other Noble gases) and Water vapour.

Although there doesn't seem to be much going on, the Earth and its crust are very dynamic.
It's just that things take such a <u>long</u> time.

STRUCTURE OF THE EARTH

The Earth is nearly spherical and has a layered structure as follows where ...

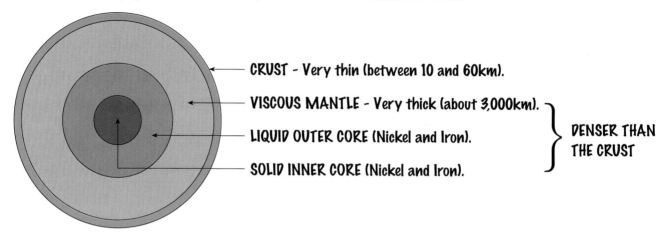

CRUST - Very thin (between 10 and 60km).

VISCOUS MANTLE - Very thick (about 3,000km).

LIQUID OUTER CORE (Nickel and Iron).

SOLID INNER CORE (Nickel and Iron).

} DENSER THAN THE CRUST

- ... the average density of the Earth is <u>MUCH GREATER</u> than ...
- ... the average density of the rocks which form the CRUST.
- This proves that the INTERIOR OF THE EARTH ...
- ... is made of a DIFFERENT and DENSER MATERIAL than that of the crust.

THE CRUST

- At the surface of the Earth sedimentary rocks exist mainly in LAYERS, ...
- ... where the younger sedimentary rocks <u>usually</u> lie on top of older rocks.

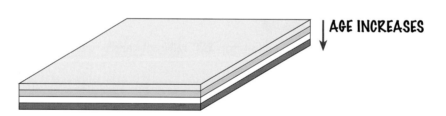

↓AGE INCREASES

- However, sedimentary rock layers are often found ...

... TILTED ...   ... FOLDED ...   ... FRACTURED ...   ... TURNED UPSIDE DOWN.

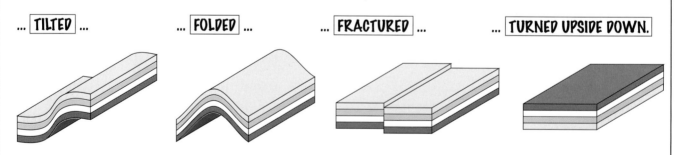

- All this shows ...
- ... that the EARTH'S CRUST HAS BEEN SUBJECTED TO VERY LARGE FORCES ...
- ... to cause this movement of the sedimentary rock layers ...
- ... and is very UNSTABLE!

- The Earth has a layered structure made up of the Crust, Viscous Mantle, Liquid Outer Core and Solid Inner Core.
- The crust is a dynamic structure that can be changed in many ways.

MOVEMENT OF THE CRUST

- The Earth's CRUST is 'cracked' into several large pieces ...
- ... called TECTONIC PLATES ...
- ... which move slowly at speeds of a few cm per year ...
- ... driven by CONVECTION CURRENTS ...
- ... in the MANTLE which are caused ...
- ... by HEAT released from RADIOACTIVE DECAY.

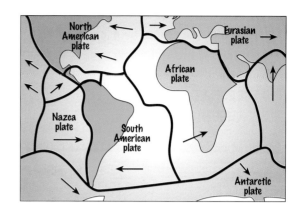

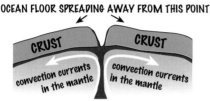

OCEAN FLOOR SPREADING AWAY FROM THIS POINT

HOT MOLTEN ROCK COMING UP TO THE
SURFACE AND SPREADING SIDEWAYS SLOWLY!

In other words,
- New crust is formed where the rising convection current reaches the crust ...
- ... and old crust disappears where the convection current starts to fall ...
- ... causing the land masses on these plates to move slowly across the globe!

Also,
- Where two land masses collide, mountain ranges are formed e.g. the Himalayas.
- These take millions of years to form and ...
- ... they replace older mountain ranges ...
- ... which have become worn down by weathering and erosion.

EVIDENCE FOR TECTONIC PLATES

Evidence for the TECTONIC PLATES theory has been gained by comparing the EAST COAST of SOUTH AMERICA and the WEST COAST of AFRICA. Although separated by thousands of kilometres of ocean, they have ...

1. CLOSELY MATCHING SHAPES

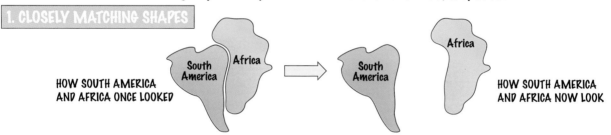

HOW SOUTH AMERICA
AND AFRICA ONCE LOOKED

HOW SOUTH AMERICA
AND AFRICA NOW LOOK

2. SIMILAR PATTERNS OF ROCKS AND FOSSILS

- ROCKS of the SAME TYPE and AGE have been found ...
- ... which contain FOSSILS of the SAME PLANTS and ANIMALS e.g. the MESOSAURUS.

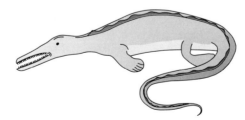

Although ...
People used to believe that the features of the Earth's surface were caused by SHRINKAGE when the Earth cooled!! We now reject this in favour of TECTONIC THEORY which explains MOUNTAIN BUILDING and the movement of the CONTINENTS from how they were (as GONDWANALAND) to what they look like today.

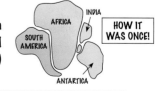

HOW IT WAS ONCE!

- The Earth's crust is cracked into several large pieces called Tectonic Plates. • Convection currents in the mantle cause these plates to move apart or collide. • Evidence for tectonic plates has been gained by comparing the shape and pattern of rocks and fossils of the East Coast of South America and the West Coast of Africa.

1. Limestone can be used as a neutralising agent, and also to make glass. Choose words from the list for each of the spaces 1 - 4 in the word equations.
   A. CARBON DIOXIDE
   B. SAND
   C. SODIUM CARBONATE
   D. CALCIUM HYDROXIDE

   CALCIUM CARBONATE $\xrightarrow{\text{HEAT}}$ CALCIUM OXIDE + _ _ _1_ _ _

   CALCIUM OXIDE $\xrightarrow{\text{WATER}}$ _ _2_ _

   LIMESTONE + _ _3_ _ + _ _4_ _ $\xrightarrow{\text{HEAT}}$ GLASS

2. The diagram below shows how oil can be found in sedimentary rock. Choose words from the list for each of the labels 1 - 4 on the diagram.
   A. WATER
   B. NON-POROUS ROCK
   C. OIL
   D. GAS

   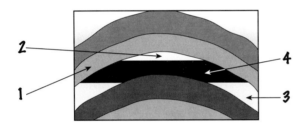

3. Marble is a metamorphic rock. Which **TWO** of the following statements are correct about marble?
   A. It is formed from molten rock which cools within the earth's crust.
   B. It is formed from small particles which fall to the bottom of shallow seas.
   C. It consists of tiny crystals which can form bands.
   D. It is formed from molten rock which cools above the earth's crust.

4. Which **TWO** of the following statements comparing oil fractions with short hydrocarbon chains, with long hydrocarbon fractions are true?
   A. Long chains are more volatile.
   B. Short chains are more viscous.
   C. Long chains are less flammable.
   D. Short chains have higher boiling points.
   E. Long chains are more viscous.

5. Global warming is mainly the result of burning fossil fuels.

5.1 Which of these substances is **NOT** a fossil fuel?
   A. Coal
   B. Oil
   C. Wood
   D. Gas

5.2 Which of these statements best describe global warming?
   A. Carbon dioxide in the atmosphere reflects heat back into space.
   B. Carbon dioxide in the atmosphere radiates heat back into space.
   C. Carbon dioxide helps to reduce the amount of heat radiated back into space.
   D. Carbon dioxide stores heat in the atmosphere.

5.3 Our atmosphere has been much the same for the past 200 million years.
   Which of the following represents the correct proportions of gases in our atmosphere?
   A. Three fifths Nitrogen, two fifths oxygen plus small amounts of other gases.
   B. Three fifths Oxygen, one fifth Nitrogen plus one fifth other gases.
   C. Four fifths oxygen, one fifth Nitrogen plus small amounts of other gases.
   D. Four fifths Nitrogen, one fifth Oxygen plus small amounts of other gases.

5.4 Burning fossil fuels can also cause acid rain. Which gases are responsible for acid rain (besides carbon dioxide?)
   A. Oxygen and Sulphur dioxide.
   B. Nitrogen oxides, and sulphur dioxide.
   C. Oxygen and Nitrogen oxides.
   D. Water vapour and Sulphur dioxide.

6. The hydrocarbon ethene can be represented by the following ...

   $$\begin{array}{ccc} H & & H \\ | & & | \\ C & = & C \\ | & & | \\ H & & H \end{array}$$

6.1 If a fuel containing carbon, hydrogen and sulphur is burnt which gases are likely to be produced?
   A. Carbon Dioxide, Hydrogen Sulphide.
   B. Carbon Dioxide, Water vapour, Sulphur Dioxide.
   C. Hydrocarbons, Sulphur Dioxide.
   D. Water vapour, Hydrogen Sulphide.

6.2 When substances burn , they react with ...
   A. Heat.
   B. Oxides.
   C. Oxygen.
   D. Carbon Dioxide.

6.3 When a substance is burned in air its elements are converted to ...
   A. Ash.
   B. Nitrates.
   C. Oxides.
   D. Carbonates.

6.4 Oil fractions which burn best ...
   A. Contain Sulphur.
   B. Have long chain hydrocarbons.
   C. Have short chain hydrocarbons.
   D. Contain Oxygen.

7. The earth's crust is 'Cracked' into several large pieces called tectonic plates which are constantly moving at speeds of a few cm. per year.

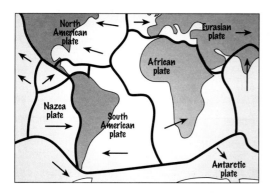

7.1 The energy which drives these plates around the globe comes from ...
   A. Ocean currents deep within the Atlantic and pacific.
   B. The cooling of the earth's core.
   C. Convection currents caused by heat from radioactive decay.
   D. The moon's gravitational pull on the earth.

7.2 The Earth is nearly spherical and has a core made of ...
   A. Nickel.
   B. Nickel and Carbon.
   C. Iron and Carbon.
   D. Nickel and Iron.

7.3 The interior of the Earth is ...
   A. The same density as the crust.
   B. Less dense than the crust.
   C. Denser than the crust.
   D. A complete mystery to everyone.

7.4 Evidence for the gradual moving apart of South America and Africa is provided by similar ...
   A. Shapes, fossils and rocks.
   B. Sizes, fossils and rocks.
   C. Shapes, and size of mountains.
   D. Shapes and species of modern animals.

ENERGY

Heat (thermal) energy is transferred from HOTTER PLACES TO COOLER PLACES by three different methods, CONDUCTION, CONVECTION and RADIATION.

CONDUCTION The key point about this type of energy transfer is that the SUBSTANCE ITSELF DOESN'T MOVE. Metals are particularly good conductors while INSULATORS (poor conductors) are also important.

Saucepans and frying pans tend to be made of metal so that heat conducts through them quickly ...

... Their handles, however, are made of wood or plastic which is a poor conductor, to prevent the heat conducting along them and making them too hot to hold.

GOOD CONDUCTORS
- ALL METALS especially COPPER and ALUMINIUM
POOR INSULATORS

REMEMBER! CONDUCTION HAPPENS MAINLY IN SOLIDS.

GOOD INSULATORS
- MOST NON-METALS
- GLASS • WOOD
- PLASTIC
- ALL GASES
POOR CONDUCTORS

CONVECTION

Liquids and gases can FLOW and can therefore transfer HEAT ENERGY from hotter to cooler areas by <u>their own movement</u>.

LIQUID or GAS
Transfer of Heat Energy

CONVECTION OF HEAT FROM AN OPEN FIRE

CONVECTION CURRENTS occur in houses which use traditional heating methods e.g. electric fires, open fires, radiators etc.

RADIATION

Energy is continually being transferred to and from all objects by radiation. Hot objects transfer energy through INFRA-RED RADIATION and the hotter the object, the more energy it radiates.

- How much radiation is <u>given out</u> or <u>taken in</u> by an object depends on its SURFACE.

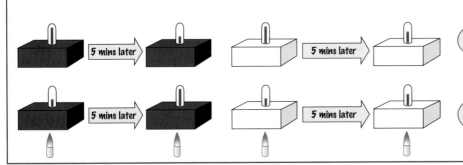

DARK MATT SURFACES EMIT MORE RADIATION THAN LIGHT SHINY SURFACES AT THE SAME TEMPERATURE

DARK MATT SURFACES ARE BETTER ABSORBERS (POORER REFLECTORS) OF RADIATION THAN LIGHT SHINY SURFACES

- Heat energy can be transferred from a hotter place to a cooler place by conduction, convection and radiation.

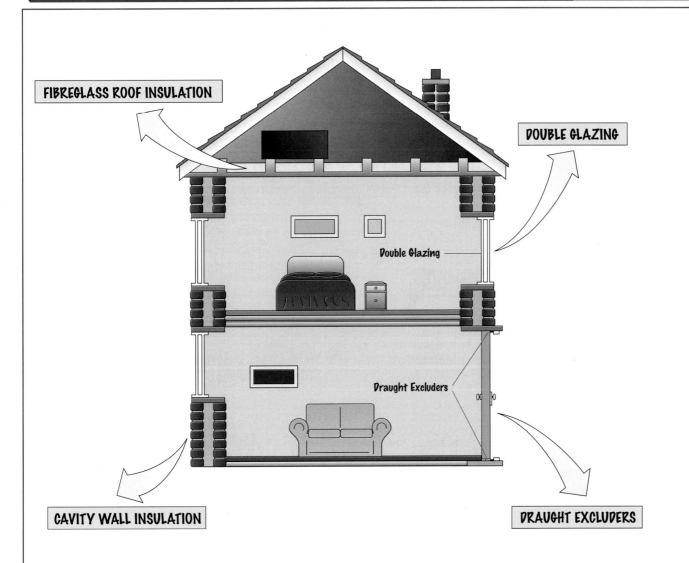

FIBREGLASS ROOF INSULATION

Reduces heat loss by conduction and convection, because of the layer of air (a good insulator) trapped between the fibres.

DOUBLE GLAZING

Reduces heat loss by conduction and convection, because of the air between the panes of glass.

CAVITY WALL INSULATION

Reduces heat loss by conduction and especially convection by trapping the air in foam.

DRAUGHT EXCLUDERS

Reduces heat loss by convection by keeping as much warm air as possible inside.

> **F♂** • Methods of saving heat energy losses from a house are: Roof Insulation, Double Glazing, Cavity Wall Insulation and Draught Excluders. • Air forms the basis of most energy saving devices.

Most of the energy transferred in homes and industry is ELECTRICAL ENERGY because ...

... it is easily transferred as HEAT, LIGHT, SOUND and MOVEMENT (Kinetic) energy.

- The rate of energy transfer is measured in WATTS. This gives the power of the appliance.
- 1 Watt is the transfer of 1 Joule of energy in 1 second.

POWER RATINGS OF SOME DOMESTIC APPLIANCES

- Most appliances in the home depend on the transfer of ELECTRICAL ENERGY into other FORMS OF ENERGY.
- All appliances have a POWER RATING which tells us how much ENERGY IS TRANSFERRED by that appliance EVERY SECOND. Here are some examples ...

ELECTRICAL ENERGY IS TRANSFERRED TO APPLIANCE BY CABLE

1000 watts (W) = 1 kilowatt (kW)

	POWER RATING (W)	POWER RATING (kW)	ENERGY TRANSFERRED PER SECOND
Computer Monitor	200 W	0.2 kW	200 J/s
Power Drill	600 W	0.6 kW	600 J/s
Food Mixer	150 W	0.15 kW	150 J/s
Toaster	900 W	0.9 kW	900 J/s

WORKING OUT THE ENERGY TRANSFERRED BY AN ELECTRICAL DEVICE

Our formula is ...   **ENERGY TRANSFERRED (J) = POWER (W) x TIME (s)**

Formula Triangle

From this, you can see that the energy transferred depends on ...

(A) How long the appliance is switched on (in seconds).

(B) How fast the appliance transfers energy (its POWER in Watts).

REMEMBER!   POWER IN <u>WATTS</u>, TIME IN <u>SECS</u> ⟶ ENERGY in <u>JOULES</u>

EXAMPLE The rock band 'Nobbler Jughead and the Asprin bottles' have just got two brand new speakers to use at their gigs. Both speakers are 700 watts. If their latest single lasts 2 mins 40 secs, how much electrical energy will have been transferred to sound energy during its performance?

Using our equation:-   Energy transferred   =   Power x time

                      "   =   700W x 160s (seconds remember!)

                      "   =   112,000 Joules

But since there are 2 speakers, the answer is 112,000 x 2 = <u>224,000 Joules!</u>

CALCULATING THE POWER RATING OF A DEVICE

EXAMPLE A toaster transfers 108,000J of electrical energy into heat in the 120 secs it takes to toast a piece of bread. Calculate it's power rating. Rearrange the formula using the formula triangle into ...

$$\text{POWER} = \frac{\text{ENERGY TRANSFERRED}}{\text{TIME}} = \frac{108,000J}{120s} = 900J/s.$$

<u>900 J/s is 900 Watts or 0.9 Kilowatts</u>

- The power rating of an appliance tells us how much energy is transferred by that appliance every second.
- Energy transferred (J) = Power (W) x Time (s) • Power (W or J/s) = Energy transferred (J) ÷ Time (s)

THE ELECTRICITY METER IN YOUR HOME

Your meter at home may show a reading like this ...

Your latest bill may look like this ...

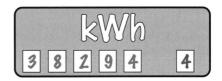

REB Regional **Electricity** Board

Customer No.	Statement Date
3 263 1319 12	26 Jan 1999
Most recent reading	38223 . 2 kWh
Previous reading	37519 . 1 kWh
Units used	704 . 1 kWh
	(Units)

704.1 units at 6.9p per unit	=	£48.58
FIXED CHARGES	=	£10.00
plus VAT @ 5%	=	£61.51

The letters kWh represent ...

... kilowatt-hours, a unit of <u>ENERGY.</u>

These are sometimes called "Units," and are a measure

of the electricity you have used.

1 kWh = 3,600,000 Joules.

THE KILOWATT-HOUR

The Kilowatt-hour is a unit of ENERGY ...

... please remember it is NOT a unit of power- that's the kilowatt!!

A electrical appliance transfers 1 kWh of energy if it transfers energy at the rate of 1 kilowatt for one hour.

A 200 watt T.V. set   ... transfers 1 kWh of energy if it is switched on for **5** hours.

A 500 watt vacuum cleaner 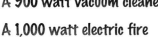   ... transfers 1 kWh of energy if it is switched on for **2** hours.

A 1,000 watt electric fire   ... transfers 1 kWh of energy if it is switched on for **1** hour.

KILOWATT-HOUR CALCULATIONS

We need to use the following formula ...

ENERGY TRANSFERRED (kWh) = POWER (kW) x TIME (h)

Formula Triangle

EXAMPLE

A 1500 watt electric hot plate is switched on for 4 hours. How much does it cost if electricity is 6p per unit?

Using the above equation ... Energy transferred (kWh) = 1.5(kW) x 4 (hour)

kilowatts remember!

= 6 kilowatt-hours (or UNITS!)

But, | **TOTAL COST = NUMBER OF UNITS x COST PER UNIT** |

Therefore Total Cost   = 6 x 6

= 36 pence

Remember, to do these calculations, you need ...

... to make sure the POWER is in KILOWATTS, AND ...

... to make sure that the TIME is in HOURS.

- An electrical appliance transfers 1kWh or 1 Unit of energy if it transfers energy at the rate of 1kW for 1 hour.
- Energy transferred (kWh) = Power (kW) x Time (h)
- Total Cost = Number of units x Cost per unit

GRAVITATIONAL POTENTIAL ENERGY

WEIGHT

WEIGHT is due to the force of GRAVITY on an object.

- Near the surface of the Earth the GRAVITATIONAL FIELD STRENGTH (g) is 10N/Kg. ...
- ... so EVERY 1KG OF MATTER experiences a DOWNWARDS FORCE or has a WEIGHT of 10N.

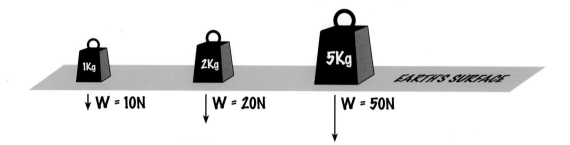

GRAVITATIONAL POTENTIAL ENERGY

- This is the ENERGY STORED in an object because of the HEIGHT ...
- ... through which the WEIGHT of the object has been LIFTED ...
- ... AGAINST the force of GRAVITY.
- IF IT CAN FALL, IT'S GOT GRAVITATIONAL POTENTIAL ENERGY.

A skier at the top of a mountain has gravitational potential energy.

USING GRAVITATIONAL POTENTIAL ENERGY TO GENERATE ELECTRICITY

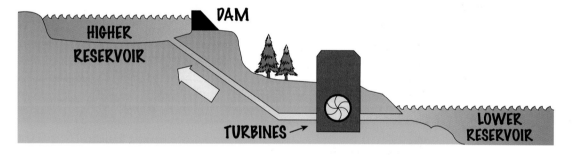

- Water is pumped up into the higher reservoir against the force of gravity using electrical energy in times of low electricity demand.

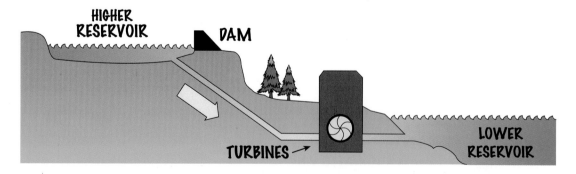

- Water is allowed to flow down into the lower reservoir converting gravitational potential energy into electrical energy in times of peak electricity demand.

- A mass of 1Kg has a weight of 10N near the surface of the Earth.
- The energy stored in an object because of the height through which the weight of the object has been lifted against gravity is called gravitational potential energy.

When electrical devices transfer energy, only part of it is USEFULLY TRANSFERRED to where it is wanted and in the form that it is wanted. The remainder is transferred in some non-useful way and is therefore 'wasted'.

USEFUL ENERGY TRANSFER　　　　　　　**'WASTE' ENERGY**

COMPUTER MONITOR

ELECTRICAL ENERGY ————→ LIGHT & SOUND　　　... HEAT

FOOD MIXER

ELECTRICAL ENERGY ————→ MOVEMENT (KINETIC)　　... HEAT and SOUND

POWER DRILL

ELECTRICAL ENERGY ————→ MOVEMENT (KINETIC)　　... HEAT and SOUND

TOASTER

ELECTRICAL ENERGY ————→ HEAT　　　... LIGHT

JET ENGINE

POTENTIAL (CHEMICAL) ————→ MOVEMENT (KINETIC)　　... HEAT, SOUND and LIGHT

EFFICIENCY OF APPLIANCES

HEAT
150 Joules/sec
"WASTED"

LIGHT
20 Joules/sec
"USEFUL"

SOUND
30 Joules/sec
"USEFUL"

ELECTRICAL
200 Joules/sec

EXAMPLES

A car engine is **20%** efficient.

A filament lamp is **15%** efficient.

A microwave is **55%** efficient.

- The fraction of the energy supplied to an appliance ...
- ... which is usefully transferred ...
- ... is called the efficiency of the appliance.

The 'wasted' energy, and the 'useful' energy are both eventually transferred to the surroundings which become WARMER. This energy eventually becomes so spread out that further energy transfer is difficult.

- When electrical appliances transfer energy only part of it is usefully transferred, the remainder is transferred in some non-useful way.
- The fraction of the energy supplied to an appliance which is usefully transferred is called the efficiency of the appliance.

ELECTRICITY FROM NON-RENEWABLE ENERGY RESOURCES

- NON RENEWABLE ENERGY RESOURCES are those that will ONE DAY RUN OUT and ...
- ... once they have been used they CANNOT BE USED AGAIN. Examples of these are ...

These three are called FOSSIL FUELS

This is NOT a FOSSIL FUEL

| COAL | OIL | GAS | NUCLEAR FUEL |

They are ALL used in the GENERATION of ELECTRICITY where ...

... THE FUEL IS USED TO GENERATE HEAT , WHICH THEN BOILS WATER TO MAKE STEAM TO DRIVE THE TURBINES , WHICH TURN THE GENERATORS PRODUCING ELECTRICITY.

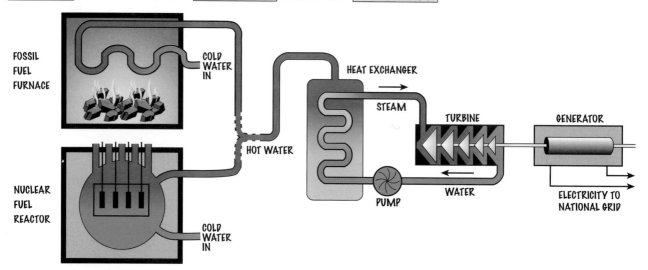

FOSSIL FUEL FURNACE

COLD WATER IN

HEAT EXCHANGER

STEAM

TURBINE

GENERATOR

HOT WATER

NUCLEAR FUEL REACTOR

COLD WATER IN

PUMP

WATER

ELECTRICITY TO NATIONAL GRID

ELECTRICITY FROM RENEWABLE ENERGY RESOURCES

- RENEWABLE ENERGY RESOURCES are those that WILL NOT RUN OUT and ...
- ... are CONTINUALLY BEING REPLACED. Examples of these are ...

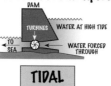

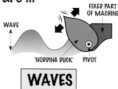

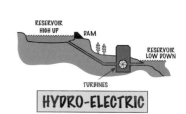

| WIND | TIDAL | WAVES | HYDRO-ELECTRIC |

- These are ALL used in the GENERATION OF ELECTRICITY where ...

- ... the ENERGY RESOURCE is used to DRIVE THE TURBINES DIRECTLY etc. ...
- ... so no nasty burning is involved. But some other renewables work differently ...

... GEOTHERMAL

- The steam needed to drive turbines can be made ...
- ... by pumping cold water through hot rocks deep in the earth's surface.
- The heat energy comes from the decay of radioactive elements e.g. Uranium.
- This happens much more slowly than in nuclear reactors.

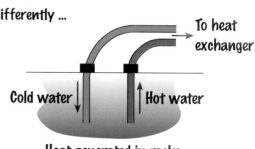

To heat exchanger

Cold water   Hot water

Heat generated in rocks

... SOLAR

- Solar cells transfer sunlight directly into electricity.
- They are expensive but if there's plenty of sunshine ...
- ... they'll provide a moderate amount of electricity.

... and WOOD

- Trees can be grown to replace trees that have been burned to provide energy for heating.

- There are two types of energy resources: Non-renewable and Renewable.

ANALYSIS OF NON-RENEWABLES

Problems, 'solutions' and characteristics.

COAL

GLOBAL WARMING DUE TO CARBON DIOXIDE.
ACID RAIN DUE TO SULPHUR DIOXIDE.

SULPHUR DIOXIDE CAN BE 'SCRUBBED' FROM THE FUMES AFTER BURNING. THIS REDUCES ACID RAIN BUT DOESN'T HELP GLOBAL WARMING.

OVER A CENTURY'S WORTH OF COAL LEFT ACCORDING TO SOME ESTIMATES. RELATIVELY CHEAP, AND SOMETIMES EASY TO OBTAIN.

SHORT START-UP TIME.
OUTPUT CAN BE REGULATED QUITE EASILY.

OIL and **GAS**

GLOBAL WARMING DUE TO CARBON DIOXIDE.
OIL SLICKS ON BEACHES FROM TANKERS AGROUND.

THE ONLY SOLUTION IS TO STOP USING OIL AND GAS.

30 PLUS YEARS LEFT ACCORDING TO SOME ESTIMATES. VARIABLE IN PRICE AND CAN BE DIFFICULT TO FIND.

SHORT START-UP TIME.
OUTPUT EASY TO REGULATE.

NUCLEAR
URANIUM + PLUTONIUM

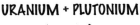

ESCAPE OF RADIOACTIVE SUBSTANCES. ELIMINATION OF RADIOACTIVE WASTE.

BUILD IN SPARSELY POPULATED AREAS AND HAVE MANY BUILT-IN SAFEGUARDS. SINK WASTE DEEP UNDERGROUND IN CONCRETE AND LEAD CHAMBERS.

THERE'S A REASONABLE AMOUNT OF URANIUM AROUND BUT IT'S DIFFICULT TO PRODUCE IN THE RIGHT QUALITY AND THE PRICE IS HIGH.

MODERN REACTORS ARE FAIRLY FLEXIBLE IN MEETING DEMAND.

You'll notice all the non-renewables on this page belt out energy but are slowly poisoning the earth.
The renewables are very "earth friendly" but can't really meet the demand.

- Coal, Oil, Gas and Nuclear are non-renewable energy resources used to generate electricity.
  Each one has its advantages and disadvantages.

ANALYSIS OF RENEWABLES

Problems, 'solutions' and characteristics.

WIND

WINDMILLS SPRINGING UP IN RURAL "BEAUTY SPOTS" (VISUAL POLLUTION).

PLACE THEM IN DESOLATE HIGH MOORLAND AREAS.

ELECTRICITY OUTPUT DEPENDS ENTIRELY ON THE PRESENCE AND STRENGTH OF THE WIND.

OUTPUT CAN'T BE GEARED TO RESPONSE UNLESS ENERGY IS STORED. EVEN THEN THE CAPACITY IS SMALL.

TIDAL and WAVE

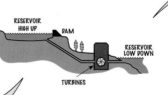

'VISUAL' POLLUTION CAUSED IN BAYS + ESTUARIES. DANGER TO SHIPPING LANES.

NO REAL SOLUTION. PERHAPS THIS IS A PRICE WORTH PAYING.

OUTPUT OF ELECTRICITY DEPENDS ON THE SIZE OF THE WAVES OR THE HEIGHT BETWEEN LOW TIDE AND HIGH TIDE.

TO MEET HIGH SPOTS IN DEMAND ENERGY HAS TO BE STORED, OR DAM WATER RELEASED WHEN DEMAND IS HIGH.

HYDRO-ELECTRIC

HABITATS MAY BE DESTROYED WHEN VALLEYS ARE FLOODED TO CREATE DAMS.

LARGE MAMMALS CAN BE MOVED ELSEWHERE BUT SMALL ANIMALS AND PLANTS ARE LOST.

AS LONG AS THERE'S ADEQUATE RAINFALL THIS IS A RELIABLE ELECTRICITY SOURCE. WATER CAN ALSO BE PUMPED INTO THE HIGHER RESERVOIR DURING OFF-PEAK TIMES.

VERY FAST RESPONSE SO CAN SUPPORT THE NATIONAL GRID IN TIMES OF EXCESSIVE DEMAND.

- • Wind, Tidal, Wave and Hydro-Electric are renewable energy resources used to generate electricity. Each one has its advantages and disadvantages.

1. Choose words from the list for each of the spaces 1 - 4 in the sentences which describe the fire (electric) used to heat a room in a house.
   A. CONDUCTION
   B. CONVECTION
   C. RADIATION
   D. INSULATION

   The on/off switch is made of plastic which provides ____1____ to prevent your hand being burnt. Heat travels through the metal casing of the fire by ____2____, but hot air rises from the fire and warms the room by ____3____. If you stand with your back against the same wall that the fire is on, thermal energy will not reach you by ____4____.

2. The amount of energy transferred by a kettle can be worked out by multiplying together which two of the following?
   A. EFFICIENCY.
   B. TIME.
   C. COST PER UNIT.
   D. POWER.
   E. NEWTONS.

3. Which <u>TWO</u> of these provide the best explanation of convection?
   A. Liquids and gases can flow.
   B. Thermal energy passes from molecule to molecule.
   C. Energy is carried from warmer to cooler places.
   D. Energy is transferred by waves.
   E. Heat travels trough a vacuum.

4. The diagram below shows a power drill which can be either used from the mains supply or has a battery attachment.

4.1 Which of the following statements accurately describes the energy transfer when the drill is plugged into the mains supply?
   A. ELECTRICAL ⟶ LIGHT + SOUND
   B. CHEMICAL ⟶ MOVEMENT + HEAT + SOUND
   C. ELECTRICAL ⟶ MOVEMENT + HEAT + SOUND
   D. CHEMICAL ⟶ MOVEMENT + HEAT + LIGHT

4.2 If the drill is attached to the battery which statement describes the energy transfer?
   A. ELECTRICAL ⟶ MOVEMENT + HEAT + SOUND
   B. CHEMICAL ⟶ MOVEMENT + HEAT + SOUND
   C. ELECTRICAL ⟶ CHEMICAL ⟶ HEAT + MOVEMENT + SOUND
   D. CHEMICAL ⟶ ELECTRICAL ⟶ MOVEMENT + HEAT + SOUND

4.3 Which of the following best describes the efficiency of an appliance?
   A. The fraction of the energy it usefully transfers.
   B. The fraction of the energy it transfers as waste energy.
   C. The fraction of the total energy released as heat.
   D. The fraction of energy it saves.

4.4 How could the drill be made more efficient?
   A. Supply it with more power.
   B. Increase the amount of useful energy it transfers.
   C. Increase the amount of energy supplied to it.
   D. Provide a higher voltage.

5. The vacuum cleaner has a power rating of 600 watts.

   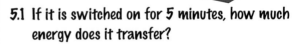

5.1 If it is switched on for 5 minutes, how much energy does it transfer?
   A. 180,000 Joules.
   B. 600 Joules.
   C. 3000 Joules.
   D. 120 Joules.

5.2 How much energy will be transferred in kilowatt-hours if the vacuum cleaner is left on for 15 minutes?
   Energy transferred = Power × time
       (kWh)   (kW)   (h)
   A. 9000 kWh.
   B. 150 kWh.
   C. 9 kWh.
   D. 0.15 kWh.

5.3 If electricity costs 12p per unit. How much will it cost to run for 1 hour?
   A. £7.20p.
   B. 0.72p.
   C. 7.2p.
   D. 12p.

5.4 What types of energy are transferred as 'waste energy' by the vacuum cleaner?
A. Light and Sound.
B. Heat and Sound.
C. Movement and Sound.
D. Movement and Heat.

6. The diagram shows a 'pumped storage' hydroelectric power station, through which water flows to turn turbines to produce electricity when needed. At times of low demand electricity is used to pump water to the higher reservoir.

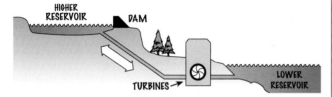

CHANGE IN GRAVITATIONAL POTENTIAL ENERGY (J)

=

WEIGHT (N) X CHANGE IN VERTICAL HEIGHT (M)

6.1 During times of high electricity demand what energy transfers take place at the power station.
A. Electrical ——————→ Heat + Kinetic
B. Gravitational ——————→ Electrical
C. Electrical ——————→ Gravitational
D. Electrical ——————→ Kinetic + Gravitational

6.2 During times of low electricity demand what energy transfers take place at the power station.
A. Electrical ——————→ Heat + Kinetic
B. Gravitational ——————→ Electrical
C. Gravitational ——————→ Electrical + Kinetic
D. Electrical ——————→ Gravitational

6.3 These types of power station are ideal to support peak electricity demand because ...
A. They have a very fast start up time.
B. They can recycle water.
C. They rely on natural rainfall.
D. They are inexpensive to construct.

6.4 Which of the following statements is __NOT__ true.
A. Hydro electric power stations have fast start up times.
B. Hydro electric power stations affect wildlife and ecosystems.
C. Hydro electric power stations produce heavy pollution.

D. Hydro electric power stations are usually in hilly or mountainous regions.

7. The diagram shows an electric kettle.

7.1 Which energy transfers take place when the kettle is switch on?
A. Electrical ——————→ Heat
B. Electrical ——————→ Heat + Sound
C. Electrical ——————→ Heat + Light
D. Electrical ——————→ Light + Sound

7.2 Some kettles have shiny silver exteriors so that ...
A. They conduct less heat to the surroundings.
B. They convect less heat to the surroundings.
C. They radiate less heat to the surroundings.
D. They reflect heat back into the water.

7.3 Only filling the kettle with as much water as you need is good for the environment because ...
A. Less steam is released into the atmosphere.
B. Less heat is radiated into the atmosphere.
C. Less heat is convected into the atmosphere.
D. Less carbon dioxide is released into the atmosphere by the power station.

7.4 We need to be careful to avoid over using electrical appliances because ...
A. They may quickly burn out.
B. Energy is precious and some power stations contribute to the greenhouse effect.
C. The demand may not be able to match the supply.
D. They radiate heat out and cause global warming.

ELECTRICITY

POTENTIAL DIFFERENCE (P.d.) AND CURRENT IN CIRCUITS

POTENTIAL DIFFERENCE AND CURRENT

An ELECTRIC CURRENT will flow through an ELECTRICAL COMPONENT (or device) ...

... if there is a VOLTAGE or POTENTIAL DIFFERENCE (p.d.) across the ends of the component.

The amount of current that flows depends on two things:

1. THE POTENTIAL DIFFERENCE (p.d) ACROSS THE COMPONENT

The BIGGER the POTENTIAL DIFFERENCE or VOLTAGE across a component ...

... the BIGGER the CURRENT that flows through the component.

(a)

Cell provides p.d. ...
... across the lamp.
A current flows and ...
... the lamp lights up.

(b)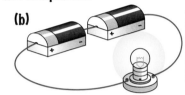

Two cells together provide ...
... a bigger p.d. across the lamp.
A bigger current now flows ...
... and the lamp lights up brighter.

2. THE RESISTANCE OF THE COMPONENT

COMPONENTS RESIST the FLOW of CURRENT THROUGH THEM. They have RESISTANCE.

The BIGGER the RESISTANCE of a COMPONENT or COMPONENTS ...

... the SMALLER the CURRENT that flows for a PARTICULAR VOLTAGE.

(c)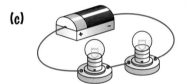

Two lamps together have a bigger resistance.
A smaller current now flows and ...
... the lamps light up dimmer (compared to (a))

... or the BIGGER the VOLTAGE needed to maintain a PARTICULAR CURRENT.

(d)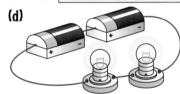

Two cells together provide a bigger p.d. ...
... and the same current as in (a) will now flow and ...
... the lamps light up brighter (compared to (c))

MEASUREMENT OF POTENTIAL DIFFERENCE AND CURRENT

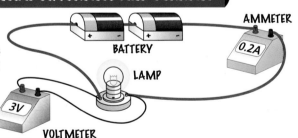

The potential difference (p.d.) across
a component in a circuit is measured
in volts (V) using a VOLTMETER
connected across the component
(in PARALLEL)

The current flowing through a
component in a circuit is measured
in amperes (A) using an AMMETER
connected in SERIES.

CURRENT-VOLTAGE GRAPHS

These graphs show how the CURRENT through a component varies with the VOLTAGE across it.

You are more likely to be asked to identify these rather than explain them, but it's better to be on the safeside!

1. RESISTOR AT CONSTANT TEMP.

Current is
proportional
to voltage.

2. A FILAMENT LAMP.

This shows how
resistance increases as
the lamp gets hot.

3. A DIODE.

A diode allows current
to flow in only one
direction.

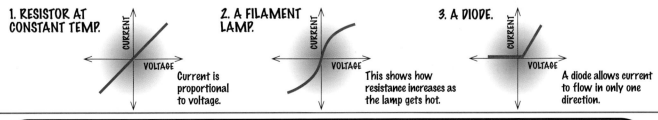

- The amount of current flowing through a component depends on the Potential Difference or Voltage across the
  component and the Resistance of the component.
- Current is measured using an Ammeter and p.d. is measured using a Voltmeter.

COMPONENTS CONNECTED IN SERIES

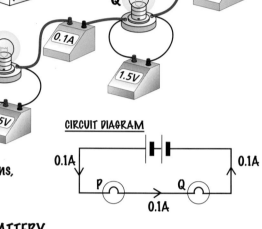

In a series circuit ...
1. THE CURRENT PASSING THROUGH ...
   ... EACH BULB (or any other component) ...
   ... is ALWAYS THE SAME.
   e.g. each ammeter reading is 0.1A.

2. Each bulb has a RESISTANCE ...
   ... and the TOTAL RESISTANCE IS EQUAL TO ...
   ... EACH INDIVIDUAL RESISTANCE ADDED TOGETHER.
   e.g. If P has a resistance of 15 ohms and Q has a resistance of 15 ohms,
   then total resistance = 15 ohms + 15 ohms = 30 ohms.

3. THE POTENTIAL DIFFERENCE (p.d.) or VOLTAGE SUPPLIED BY THE BATTERY ...
   ... is DIVIDED UP BETWEEN THE TWO COMPONENTS IN THE CIRCUIT.
   e.g. the p.d. across P is 1.5V and Q is 1.5V which added together gives us the p.d. of the battery, 3V.
   In our circuit both bulbs have the same resistance and the voltage is divided equally ...
   ... but if each bulb had a different resistance then the voltage would be divided differently e.g. 2V and 1V.

COMPONENTS CONNECTED IN PARALLEL

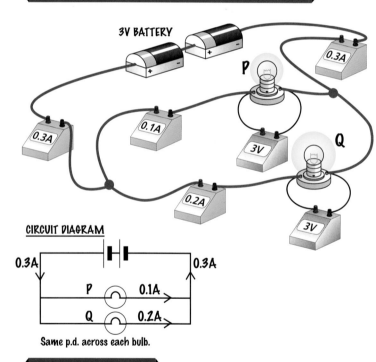

CIRCUIT DIAGRAM

Same p.d. across each bulb.

In a parallel circuit ...
1. The SUM OF THE CURRENTS ...
   ... PASSING THROUGH EACH BULB is ...
   ... EQUAL to the TOTAL CURRENT in the circuit.
   e.g. 0.3A = 0.1A + 0.2A = 0.3A

2. The AMOUNT OF CURRENT which PASSES ...
   ... THROUGH EACH BULB depends on the ...
   ... RESISTANCE OF EACH BULB.
   Bulb P has a GREATER RESISTANCE than bulb Q ...
   ... so only 0.1 amp PASSES THROUGH bulb P.
   while 0.2 amps pass through bulb Q.

3. The POTENTIAL DIFFERENCE ACROSS.
   ... EACH BULB is the SAME.
   e.g. each bulb has a p.d. of 3V across it.

CELLS IN SERIES

The TOTAL POTENTIAL DIFFERENCE provided by cells ...
... CONNECTED in SERIES is the SUM of the P.D. ...
... of EACH CELL SEPARATELY providing that they ...
... have been connected in the same direction.

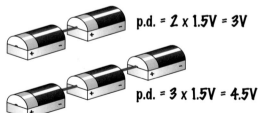

p.d. = 2 x 1.5V = 3V

p.d. = 3 x 1.5V = 4.5V

- Components in series have the same current flowing through them but the total voltage is divided up. Also the total resistance is equal to each individual resistance added together.
- Components in parallel have the same voltage across them but the total current is divided up.
- The total p.d. of cells connected in series is the sum of the p.d.'s of each individual cell.

WHAT IS STATIC ELECTRICITY

- TWO MATERIALS can become ELECTRICALLY CHARGED when they are RUBBED AGAINST EACH OTHER.
- The materials have become charged with STATIC ELECTRICITY meaning ...
- ... that the electricity stays on the material and doesn't move.

You can 'generate' static electricity by rubbing a balloon against a jumper.

The electrically charged balloon will then attract very small objects.

Small pieces of paper

WHY MATERIALS BECOME ELECTRICALLY CHARGED

Electric charge (static) builds up when ELECTRONS (which have a NEGATIVE charge) are "rubbed off" one material onto another.
The material receiving electrons becomes NEGATIVELY CHARGED, and the one giving up electrons becomes EQUALLY POSITIVELY CHARGED.

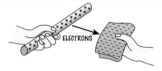

PERSPEX ROD RUBBED WITH A CLOTH

ELECTRONS

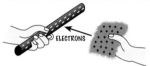

EBONITE ROD RUBBED WITH FUR

ELECTRONS

Perspex LOSES electrons ...
... POSITIVELY CHARGED.

Cloth GAINS electrons ...
... NEGATIVELY CHARGED.

Ebonite GAINS electrons ...
... NEGATIVELY CHARGED.

Fur LOSES electrons ...
... POSITIVELY CHARGED.

REPULSION AND ATTRACTION BETWEEN CHARGED MATERIALS

Very simply ... TWO MATERIALS WITH THE <u>SAME CHARGE</u> ON THEM, <u>REPEL</u> EACH OTHER while ...
     ... TWO MATERIALS WITH <u>DIFFERENT CHARGE</u> ON THEM, <u>ATTRACT</u> EACH OTHER.

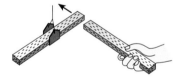

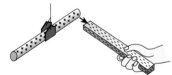

- The SUSPENDED POLYTHENE ROD ...
- ... is REPELLED by the HELD POLYTHENE ROD.

NB. WE WOULD GET THE SAME WITH PERSPEX RODS.

- The SUSPENDED PERSPEX ROD ...
- ... is ATTRACTED by the HELD POLYTHENE ROD.

NB. WE WOULD GET THE SAME IF THE RODS WERE THE OTHER WAY AROUND.

DISCHARGE

A charged conductor can be DISCHARGED i.e. have any charge on it removed by connecting it to EARTH with a CONDUCTOR.

Negatively charged dome of Van De Graff generator

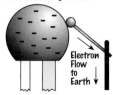

Electron Flow to Earth ↓

In this case electrons flow from the dome to Earth via the conductor ... until the dome is completely discharged.

Positively charged dome

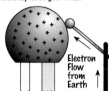

Electron Flow from Earth ↑

This time the electrons flow from Earth to cancel out the positive charge on the dome ... until the dome is completely discharged.

This flow of electrons through a solid conductor is an electric current and ...
... metals conduct electricity well because electrons from their atoms can move freely throughout the metal structure.

- The movement of electrons from one material to another creates electrically charged materials.
- Materials with like charges repel each other while materials with unlike charges attract.
- A flow of electrons is an electric current.

USING STATIC IN EVERYDAY LIFE

APPLYING PAINT EVENLY

The panel being sprayed is made positively charged, ...

... and is sprayed with negatively charged paint.

The paint particles repel each other, but ...

... are attracted to the area being painted.

This causes the paint to be applied evenly.

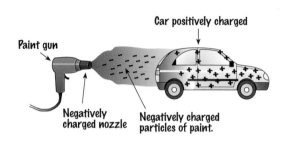

Car positively charged

Paint gun

Negatively charged nozzle

Negatively charged particles of paint.

DISCHARGING UNSAFE STATIC

FILLING AIRCRAFT FUEL TANKS

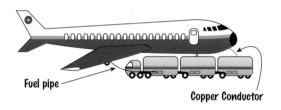

Fuel pipe

Copper Conductor

During refuelling the fuel gains electrons from the fuel pipe ...

... making the pipe positively charged and the fuel negatively charged.

The resulting voltage between the two can cause a spark ...

... (DISCHARGE). You can imagine the rest!!!

In order to prevent this, the charge must be allowed to ...

... leak away to earth via a good conductor (a copper wire).

Alternatively the tanker and plane can be connected by a conductor.

ELECTROLYSIS

- Some chemical compounds will conduct electricity ...
- ... when they are MELTED or DISSOLVED IN WATER.
- These compounds contain NEGATIVE and POSITIVE IONS ...
- ... and the electric current is due to ...
- ... <u>NEGATIVELY CHARGED IONS MOVING TO THE POSITIVE ELECTRODE</u> ...
- ... and <u>POSITIVELY CHARGED IONS MOVING TO THE NEGATIVE ELECTRODE.</u>

Battery

Positive Electrode

Negative Electrode

CHLORINE GAS RELEASED

Cl^- Cu^{++}

COPPER DEPOSITED

COPPER CHLORIDE SOLUTION

When this happens SIMPLER SUBSTANCES are released at the <u>TWO ELECTRODES.</u>

THE QUANTITY OF THESE SIMPLER SUBSTANCES RELEASED INCREASES IF ...	... THE CURRENT INCREASES ...	i.e. THE AMOUNT OF CHARGE THAT FLOWS INCREASES.
	... THE TIME FOR WHICH THE CURRENT FLOWS INCREASES.	

- A build up of static electricity can be very dangerous if discharge occurs.
- The quantity of substances released at the two electrodes during electrolysis is increased by increasing the current that flows or increasing the time for which the current flows.

Most electrical appliances are connected to the MAINS ELECTRICITY SUPPLY ...

... using a CABLE and a 3-PIN PLUG.

In the U.K. the mains supply has a VOLTAGE OF ABOUT 230 VOLTS ...

... which if it is not used safely can kill!!

MAINS SUPPLY BY SOCKET

3-PIN PLUG

CABLE

A typical appliance - a kettle

3 - PIN PLUG

EARTH WIRE (Green & Yellow)

- All appliances with outer metal cases ...
- ... are earthed.

NEUTRAL WIRE (Blue)

- Carries current away from appliance.

CABLE GRIP

FUSE

- Always part of the live circuit.
- Should be of the proper current rating.

CASING

- Plastic or Rubber because both are good insulators.

LIVE WIRE (Brown)

- Carries current to appliance.

5A

- Inner cores of COPPER ...
- ... because it's a good conductor.
- Outer layers of FLEXIBLE PLASTIC ...
- ... because it's a good insulator

CABLE

PINS

- ... while the plug has ...
- ... PINS made from BRASS ...
- ... because it's a good conductor.

WIRES   CABLE

CASING

ERRORS IN WIRING PLUGS

It is very important that all plugs are wired correctly ...

... with NO errors, for our own safety.

Below are five examples of dangerously wired plugs!

Silver foil

Bare wires showing

Proper fuse not installed

Earth wire not connected

Live and neutral wrong way round

Cable grip loose

- The mains electricity supply in the U.K. has a voltage of about 230V.
- A 3-pin plug consists of: Live wire, Neutral wire, Earth wire, Fuse, Cable grip and Casing.
- Materials found in a 3-pin plug are either good electrical conductors or good electrical insulators.

DIRECT CURRENT

Direct current (d.c) ...
- Current always flows ...
- ... in the same direction.
- Cells and batteries are d.c.

ALTERNATING CURRENT

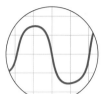

Alternating current (a.c) ...
- Current changes direction of flow ...
- ... back and forth continuously.
- Mains electricity is a.c ...
- ... of frequency 50 Hz ...
- ... i.e. no. of cycles every second.

FUSES

Very simply ...
- A FUSE is a SHORT, THIN piece of WIRE ...
- ... with a LOW MELTING POINT.

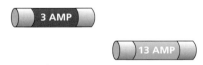

- When the CURRENT passing through it EXCEEDS ...
- ... the CURRENT RATING of the fuse, ...
- ... the fuse wire gets HOT and MELTS or BREAKS.

Too much current →

Fuse 'burns out' or melts

- This PREVENTS DAMAGE to CABLE or APPLIANCE through the possibility of OVERHEATING.

CURRENT LARGER THAN CURRENT RATING OF FUSE	→	FUSE BURNS OUT	→	CIRCUIT IS BROKEN	→	NO CURRENT FLOWS	→	CABLE OR APPLIANCE IS PROTECTED

However ...
- For this safety system to work properly the CURRENT RATING of the fuse ...
- ... must be JUST ABOVE THE NORMAL WORKING CURRENT of the appliance (see P.65).

CIRCUIT BREAKERS

Most modern houses and appliances tend to have CIRCUIT BREAKERS rather than rely on fuses. (They're sometimes called MCB's - where the 'M' stands for miniature!) They depend on an ELECTROMAGNET, which separates a PAIR OF CONTACTS, <u>WHEN THE CURRENT BECOMES HIGH ENOUGH</u>.

They work MORE QUICKLY THAN A FUSE, ...

... and are EASILY RESET by pressing a button (see P.67).

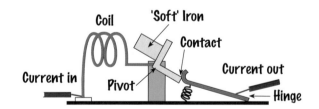

Coil   'Soft' Iron
        Contact
Current in   Pivot   Current out
        Hinge

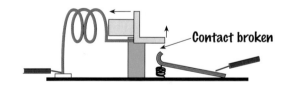

Contact broken

- Current flowing in the same direction is d.c. while current that constantly changes its direction of flow is a.c.
- A fuse and a circuit breaker are safety devices used to protect a cable or an appliance.

POWER AND ENERGY TRANSFER

An electric current is a <u>flow of charge</u> which <u>transfers energy</u> from the <u>battery or power supply</u> to the <u>components in the circuit</u>. If the component is a <u>resistor</u>, <u>electrical energy</u> is <u>transferred as heat</u>.

The <u>rate</u> of this energy transfer is the <u>POWER</u> in Joules/second or <u>WATTS</u> (W)

POWER (W) = POTENTIAL DIFFERENCE (V) x CURRENT (A)

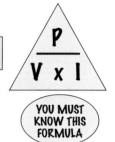

(1 watt is the transfer of 1 Joule of energy in 1 second).

YOU MUST KNOW THIS FORMULA

Appliance rating plates give us two of these so we can work out the current, and therefore the fuse we should use.

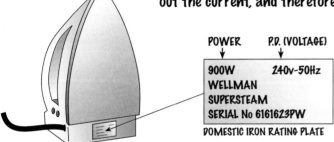

POWER   P.D. (VOLTAGE)

900W	240v-50Hz
WELLMAN	
SUPERSTEAM	
SERIAL No 6161623PW	

DOMESTIC IRON RATING PLATE

$$\text{Current} = \frac{\text{Power}}{\text{Potential Difference}}$$

$$I = \frac{900 \text{ W}}{240 \text{ W}}$$

I = <u>3.75amps, therefore we need a 5a fuse</u>

The current rating of the fuse should be as close as possible to (but <u>HIGHER</u> than) the normal current which flows through the appliance. (See P.64)

STANDARD ELECTRICAL SYMBOLS

The following standard symbols should be known. You may be asked to interpret and/or draw circuits using the following standard symbols.

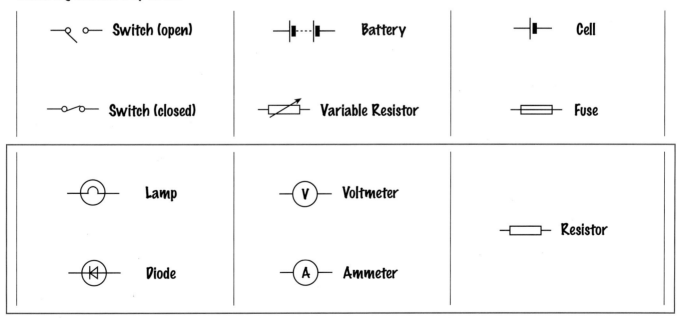

Switch (open)	Battery	Cell
Switch (closed)	Variable Resistor	Fuse
Lamp	Voltmeter	
Diode	Ammeter	Resistor

- Rate of energy transfer is called power. • Power (W) = Potential Difference (V) x Current (A).
- The current rating of a fuse should be as close as possible to, but higher than, the normal current which flows through an appliance.

MAGNETS

If a **MAGNET** is allowed to move freely ...
- ... the end of the magnet which points **NORTH** is called the **NORTH SEEKING POLE** ...
- ... and the end which points **SOUTH** is called the **SOUTH SEEKING POLE**.

ALL MAGNETS have a **REGION OF SPACE AROUND THEM** ...
- ... called the **MAGNETIC FIELD** ...
- ... which exerts a **FORCE** ...
- ... on **ANOTHER MAGNET** or **MAGNETIC MATERIAL** which enters it ...

These are called FIELD LINES.

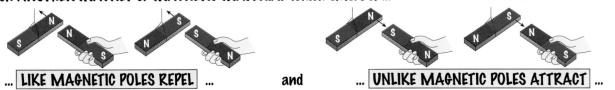

... **LIKE MAGNETIC POLES REPEL** ...   and   ... **UNLIKE MAGNETIC POLES ATTRACT** ...

- However, the magnet will only exert an **ATTRACTIVE FORCE** when an **IRON** or **STEEL BAR** enters its **MAGNETIC FIELD**.

MAGNET   IRON BAR (or STEEL)   THE MAGNET IS MOVED TOWARDS THE IRON BAR

ELECTROMAGNETS

- If an **ELECTRIC CURRENT** flows through a **COIL OF WIRE** a **MAGNETIC FIELD** is **FORMED AROUND THE COIL** ...
- ... creating an **ELECTROMAGNET** which is a **MAGNET** that can be **SWITCHED** **ON** and **OFF**.

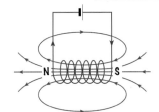

- The **MAGNETIC FIELD** formed is ...
- ... very **SIMILAR** to a **BAR MAGNET** ...
- ... and so ...
- ... **ONE END OF THE COIL BECOMES A NORTH POLE** and ...
- ... the **OTHER END BECOMES A SOUTH POLE**.
- Reversing the current, reverses the **POLES** of the electromagnet.

The strength of an electromagnet can be increased in **3 ways:-**

1. INCREASE THE NUMBER OF TURNS ON THE COIL.	2. INCREASE THE SIZE OF THE CURRENT.	3. PLACE A SOFT IRON CORE INSIDE THE COIL.
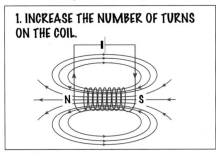	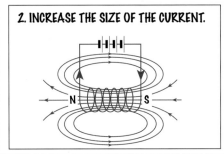	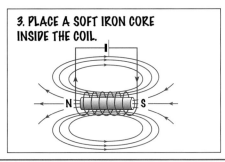

THE MOTOR EFFECT

- When a wire carrying an **ELECTRIC CURRENT** is placed in a **MAGNETIC FIELD** ...
- ... it will experience a **FORCE** causing it to **MOVE**.

Movement upwards

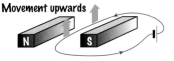

- The direction of the force on the wire can be reversed by ...
- ... **REVERSING THE DIRECTION OF FLOW OF THE CURRENT** (turn your cell around).
- ... or **REVERSING THE DIRECTION OF THE MAGNETIC FIELD** (swap your magnets).

The size of the force on the wire can be increased in **2 ways:-**

1. INCREASE THE STRENGTH OF THE MAGNETIC FIELD ... ... have stronger magnets.	2. INCREASE THE SIZE OF THE CURRENT ... ... have more cells.
	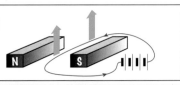

- Like magnetic poles repel and vice versa. • An electromagnet is formed when an electric current flows through a coil of wire • A wire carrying an electric current placed in a magnetic field will experience a force.

Electromagnets unlike ordinary magnets can be switched ON and OFF very quickly if need be. This makes them particularly handy for causing movement in various everyday devices.

THE LOUDSPEAKER

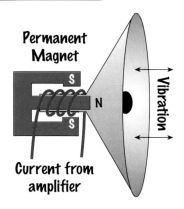

Permanent Magnet

Vibration

Current from amplifier

- Varying ALTERNATING CURRENT from the AMPLIFIER ...
- ... causes the coil to vibrate back and forth over the North pole of the permanent magnet ...
- ... which causes the cardboard cone to vibrate ...
- ... creating SOUND WAVES.

THE CIRCUIT BREAKER

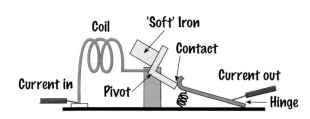

Coil   'Soft' Iron
Contact
Current out
Current in   Pivot   Hinge

- If current gets too HIGH ...
- ... then strength of field generated by the COIL INCREASES ...
- ... until it is strong enough to ATTRACT the SOFT IRON rocker ...
- ... which causes the CONTACT TO BREAK ...
- ... SWITCHING OFF the current.

THE RELAY

A RELAY is a SWITCH where a SMALL CURRENT SWITCHES ON A LARGER CURRENT (FOR SAFETY).

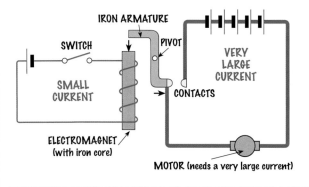

IRON ARMATURE
SWITCH   PIVOT
VERY LARGE CURRENT
SMALL CURRENT
CONTACTS
ELECTROMAGNET (with iron core)
MOTOR (needs a very large current)

- When the SWITCH is CLOSED, a SMALL CURRENT flows and the ELECTROMAGNET is SWITCHED ON.
- The PIVOTED IRON ARMATURE is PULLED TOWARDS the ELECTROMAGNET ...
- ... causing the CONTACTS TO CLOSE.
- The second circuit is now complete and a LARGE CURRENT CAN NOW FLOW.

THE DIRECT CURRENT (d.c) MOTOR

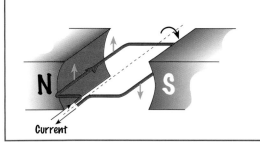

N   S
Current

Uses the principle of the MOTOR EFFECT.
- As current flows through the coil on BOTH SIDES, ...
- ... each creates a MAGNETIC FIELD ...
- ... which INTERACTS with the ...
- ... PERMANENT MAGNETIC FIELD of the MAGNET ...
- ... creating a FORCE on both sides of the COIL
- ... which ROTATES THE COIL.

- The direct current (d.c.) motor uses the principle of the motor effect to produce rotation of a coil.

MAKING ELECTRICITY BY ELECTROMAGNETIC INDUCTION

- If you move a wire, a coil of wire or a magnet ...
- ... so that the wire cuts through a magnetic field (the lines of force) ...
- ... then a voltage is induced between the ends of the wire ...
- ... and a <u>current</u> will be <u>induced</u> in the wire ...
- ... if it is part of a complete circuit.

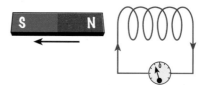

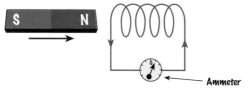

- ... while a current can be induced in the opposite direction by moving the magnet out of the coil or ...

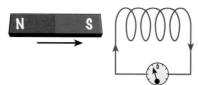

- Moving the magnet into the coil induces a current in one direction ...

- ... by moving the other pole of the magnet into the coil.

N.B. WHEN THERE'S NO MOVEMENT, THERE'S NO CURRENT.

- Generators use this principle for generating electricity by ...
- ... ROTATING A COIL OF WIRE WITHIN A MAGNETIC FIELD ...
- ... or ROTATING A MAGNET INSIDE A COIL OF WIRE.

Both of these involve a magnetic field being cut by the coil of wire ... creating an induced voltage.

INCREASING THE SIZE OF THE INDUCED VOLTAGE

1. Increase speed of movement of magnet towards coil ...	2. Increase strength of magnetic field.	3. Increase number of turns on coil.
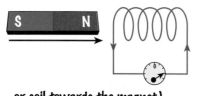	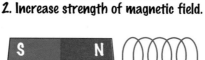 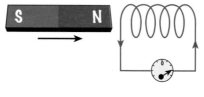	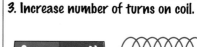 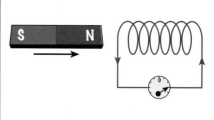
(... or coil towards the magnet.)		

TRANSFORMERS

- Transformers are used to CHANGE the VOLTAGE of an A.C. SUPPLY, UP or DOWN.
- At the POWER STATION electricity is generated and transformers are used to 'STEP-UP' this voltage ...
- ... before transmission through the NATIONAL GRID to where it is needed.
- Local transformers are then used to 'STEP-DOWN' this voltage to safe levels for use by consumers.

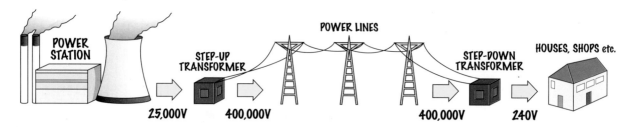

- A voltage is induced between the ends of a wire and a current is induced in the wire if it is part of a complete circuit when it cuts through a magnetic field.
- Transformers are used to change the voltage of an a.c. supply up or down.

1. Look at the circuit diagram and choose words from the list to complete the labels.
   A. LAMP.
   B. VOLTMETER.
   C. CELL.
   D. BATTERY.

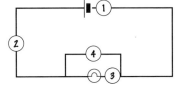

The____1____transfers energy to other components in the circuit. The____4____transfers electrical energy as light and heat. The____3____measures the p.d. across components in a circuit. The____2____measures the current flowing in the circuit.

2. The diagram shows a T.V. set. Choose words from the list for each of the spaces in the sentences.
   A. CURRENT.
   B. VOLTAGE.
   C. RESISTANCE.
   D. HEAT.

To make the T.V. set work, a____1____must be applied across it so that a____2____flows through it. Electricity doesn't flow easily through the T.V. set because it has____3____. This causes some of the electrical energy to be transferred as ____4____.

3. The diagram shows two polythene rods which are repelling each other. Which two of the following statements relating to this diagram are true?

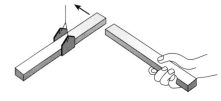

   A. One rod is negatively charged and the other is positively charged.
   B. Neither rod is charged.
   C. Materials with the same charge repel each other.
   D. These rods are made of different materials.
   E. These rods both carry the same type of charge.

4. Look at the circuit diagram below.

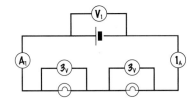

4.1 What reading would show on the ammeter A_1?
   A. 3A.
   B. 1A.
   C. 2A.
   D. 6A.

4.2 What reading would show on the voltmeter V_1?
   A. 3V.
   B. V.
   C. 6V.
   D. 7V.

4.3 What does the symbol ▌ mean in the circuit?
   A. Fuse.
   B. Switch.
   C. Cell.
   D. Resistor.

4.4 What is the power of each of the bulbs?
   A. 100 Watt.
   B. 60 Watt.
   C. 6 Watt.
   D. 3 Watt.

5. As a fuel tanker fills up a jet plane, the fuel becomes highly negatively charged and the fuel pipe positively charged?

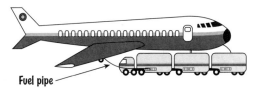

Fuel pipe

   A. Because electrons are flowing from the fuel to earth.
   B. Because electrons are transferred from the pipe to the fuel.
   C. Because electrons are transferred from the fuel to the pipe.
   D. Because of electromagnetic induction.

5.2 Why could this situation cause a serious accident.
   A. The fuel may spill from the tanker.
   B. Discharge may occur in the form of a spark.
   C. Negatively charged fuel may catch fire.
   D. The fuel pipe becomes hot because of the positive charge.

5.3 The chances of an accident can be reduced by ...
A. Earthing the fuel tank.
B. Only half filling the tank.
C. Refuelling more slowly.
D. Connect a copper wire from the fuel pipe to the fuel tanker.

5.4 Which of the following is not an example of static electricity?.
A. Sparks caused by removing a sweater.
B. A shock from a car door.
C. A shock from an electrical socket.
D. A comb picking up small pieces of paper.

6. Most electrical appliances are connected to the mains by a 3 pin plug.

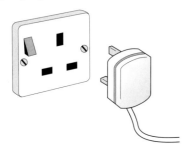

6.1 The single pin at the top of the plug is ...
A. The earth connection.
B. The live connection.
C. The neutral connection.
D. The fuse connection.

6.2 The fuse is connected to ...
A. The earth wire.
B. The live wire.
C. The neutral wire.
D. The fuse wire.

6.3 The casing and cable of the plug should be ...
A. Good conductors.
B. Made of metal.
C. Good insulators.
D. Connected to the fuse.

6.4 The U.K. mains supply is ...
A. 230 Volts and 20Hz.
B. 23 Volts and 500Hz.
C. 230 Volts and 50Hz.
D. 23 Volts and 50Hz.

7. The diagram below shows a simple direct current motor.

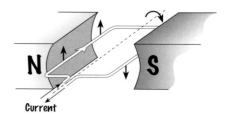

Current

7.1 When the motor is connected to a D.C. supply, ...
A. The coil spins due to the build up of static charge.
B. Each side of the coil produces a magnetic field which interacts with the permanent magnet.
C. Each side of the coil produces a magnetic field which attracts the other side.
D. Each side of the coil produces a magnetic field which repels the other side.

7.2 Which one of the following will increase the size of the force on the wires of the coil.
A. Wrap them around a heavy plastic core.
B. Wrap them around a light plastic core.
C. Reduce the current in the coil.
D. Increase the strength of the permanent magnets.

7.3 What would happen if an A.C. supply was used instead of a D.C?
A. The motor would reverse its direction of spin.
B. The motor would vibrate about a central position.
C. The motor would turn more quickly.
D. The motor would turn more slowly.

7.4 Electromagnets are used in many everyday appliances. Which of the following would NOT increase the strength of an electromagnet?
A. Increasing the number of turns on the coil.
B. Increasing the size of the current.
C. Placing a soft iron core inside the coil.
D. Increasing the frequency of supply.

INDEX

Acid 34
Acid Rain 43
Alcohol 25
Alkali 34
Alloys 28
Alternating Current Generator 68
Aluminium 28
Alveoli 9
Alveolus 8
Amino Acids 8
Ammeter 59
Anus 7
Arteries 11, 13
Atmosphere 43
Atria 12
Attraction 61

Bacteria 14
Blast Furnace 31
Blood 11
Brain 23
Bronchioles 9
Bronchus 9

Calcium Carbonate 37
Calcium Hydroxide 37
Calcium Oxide 37
Capillaries 11, 13
Carbohydrase 7, 8
Carbohydrates 8
Carbon Dioxide 19
Cell Membrane 6, 17
Cells 6, 17, 60
Cellulose Cell Wall 17
Chloroplasts 17
Ciliary Muscle 23
Circuit Breaker 64, 67
Circulatory System 11
Coal 40, 54, 55
Conduction 48
Conductor 61
Convection 48
Copper 28
Core 44
Cornea 23
Cracking 41
Crude Oil 40
Crust 44, 45

Cryolite 32
Current 59, 60, 65
Current, Alternating 64
Current, Direct 64
Current-Voltage Graphs 59
Cytoplasm 6, 17

Defence Mechanisms 14
Diabetes 24
Diaphragm 9
Diffusion 6, 9
Digestive System 7
Direct Current (d.c.) Motor 67
Discharge 61
Displacement Reaction 30
Domestic Appliances 50
Double Circulation 12

Earth 44
Earth Wire 63
Efficiency 53
Electrical Energy 50
Electrical Symbols 65
Electricity 54
Electricity Meter 51
Electrolysis 31, 32, 62
Electromagnetic Induction 68
Electromagnets 66
Energy 10
Energy Resources 54, 55, 56
Energy Transfer 50, 51, 53, 65
Excretion 5
Extraction of Aluminium 32
Eye 23

Fats 8
Fatty Acids 8
Flaccid 21
Focus 23
Fossil Fuel 40, 42
Fossils 45
Fractional Distillation 41
Fractionating Column 41
Fuse 63, 64

Gall Bladder 7
Gas 55
Generators 54

Geothermal 54
Glandular 7
Global Warming 42
Glucose 19
Glycerol 8
Gravitational Potential Energy 52
Growth 5
Guard Cells 17

Haemoglobin 11
Heart 11, 12
Heating With Carbon 31
Homeostasis 24
Hormones 22, 24
Hydro-Electric 54, 56
Hydrocarbons 40, 41

Igneous Rocks, Extrusive 39
Igneous Rocks, Intrusive 39
Indicators 33
Induced Voltage 68
Infra-Red Radiation 48
Insulation 49
Iris 23
Iron 28

Kilowatt-Hours 51

Lactic Acid 10
Large Intestine (Colon) 7
Leaf 20
Lens 23
Light Energy 19
Limestone 37
Lipase 7, 8
Live Wire 63
Liver 7
Loudspeaker 67
Lung 9

M"s", The 10
Magma 39
Magnetic Field 66
Magnets 66
Mains Electricity Supply 63
Mantle 44, 45
Metal Oxides 34
Metals 28

INDEX

Methane 42
Motor Effect 66
Mouth 7
Movement 5

Natural Gas 54
Nerve Impulses 23
Neutral Wire 63
Neutralisation 33
Neurons 23
Non-Metal Oxides 34
Non-Metals 28
Non-Renewables 54, 55
Nuclear 55
Nuclear Fuel 54
Nucleus 6, 17
Nutrition 5

Oil 54, 55
Optic Nerve 23
Ore 31
Organ 5
Organ Systems 5
Organism 5
Oxygen 19
Oxyhaemoglobin 11

Paired Spinal Nerves 23
Pancreas 7
Parallel Circuits 60
pH Scale 33
Phloem Tissue 20
Photosynthesis 19
Photosynthesis, Rate Of 19
Plant Responses 22
Plasma 11
Plastics 41
Platelets 11
Poly(ethene) 41
Potential Difference 59, 60, 65
Power 50, 51, 65
Power Ratings 50
Protease 7, 8
Proteins 8
Pupil 23
Purification Of Copper 32
PVC 41

Radiation 48
Reactivity Series 20, 31, 32
Receptors 23
Rectum 7
Red Blood Cells 11
Reduction 31
Relay 67
Renewables 54, 56
Reproduction 5
Repulsion 61
Resistance 59, 60
Respiration 5
Respiration, Aerobic 10
Respiration, Anaerobic 10
Retina 23
Rib Muscles 9
Ribs 9
Rock Cycle 39
Rock Types 38
Rocks, Igneous 38
Rocks, Metamorphic 38, 39
Rocks, Sedimentary 38, 39
Root Hair Cell 17, 20
Roots 18

Sclera 23
Sensitivity 5
Series Circuits 60
Small Intestine (Ileum) 7
Solar 54
Solvents 25
Specialised Animal Cells 6
Spinal Cord 23
Static 62
Static Electricity 61
Stem 18, 20
Stimuli 23
Stomach 7
Sugars 8
Support 21
Suspensory Ligament 23

Tectonic Plates 45
Temperature 48
Thermal Energy Transfer 48
Thorax (Chest Cavity) 9
Three-Pin Plug 63
Tidal 54, 56

Tissue 5
Tobacco 25
Toxins 14
Trachea 9
Transpiration 20, 21
Turbines 65
Turgid 21

Units 51
Universal Indicator 33

Vacuole 17
Valves 12
Veins 11, 13
Ventricles 12
Viruses 14
Voltmeter 59

Wasted Energy 53
Water 19
Water Loss 21
Wave 54, 56
Waxy Layer 20
Weight 52
White Blood Cells 11
Wind 54, 56
Wood 54

Xylem 17, 20
Xylem Tissue 20